# What Is Science?

A Guide for Those Who Love It,
Hate It, or Fear It

# What Is Science?

## A Guide for Those Who Love It, Hate It, or Fear It

## Elof Axel Carlson

*Indiana University, USA*

**World Scientific**

NEW JERSEY · LONDON · SINGAPORE · BEIJING · SHANGHAI · HONG KONG · TAIPEI · CHENNAI · TOKYO

*Published by*

World Scientific Publishing Co. Pte. Ltd.

5 Toh Tuck Link, Singapore 596224

*USA office:* 27 Warren Street, Suite 401-402, Hackensack, NJ 07601

*UK office:* 57 Shelton Street, Covent Garden, London WC2H 9HE

**British Library Cataloguing-in-Publication Data**
A catalogue record for this book is available from the British Library.

**WHAT IS SCIENCE?**
**A Guide for Those Who Love It, Hate It, or Fear It**

ISBN 978-981-122-871-1 (hardcover)
ISBN 978-981-123-010-3 (paperback)
ISBN 978-981-122-872-8 (ebook for institutions)
ISBN 978-981-122-873-5 (ebook for individuals)

For any available supplementary material, please visit
https://www.worldscientific.com/worldscibooks/10.1142/12055#t=suppl

Printed in Singapore

# Dedication

To the American Museum of Natural History in New York City for inspiring me in my youth that science would open the universe to my delight and education

# Preface

We live in a science-saturated civilization. Almost everything that enters our lives is touched by science. This is certainly true for our life-expectancy, the size of our family, the availability and quality of foods we eat, and the health sciences provided to us. We are assured the water we use will not give us dysentery. We feel confident about the prevention or treatment of most of our diseases. Science provides the clothes we wear and the homes we live in. It gives us the communication technologies that allow us to keep in contact with our scattered family. Science provides the jet travel to reach them. It offers the use of computers as word processors to compose our letters chapters, and memos. Without science we would be lucky to be alive.

Many of my outlooks on science were shaped by my being mentored by H. J. Muller (1890–1967) at Indiana University in 1953-1958 for my PhD. Muller thought deeply about science and society and frequently found himself criticized and threatened for his views against racism, sexism, spurious elitism, concern for protecting the public from radiation exposure, concern for the genetic load in the human population, and offering a program of germinal choice as a compensation for the "load of mutations" that civilization imposes on our species.

However, much of the civilized world fears science. It sees science as the source of pollution. Its role in wars threaten the lives of civilians and may potentially destroy most of human life. Science is seen as hostile to values, religion, and the arts and humanities. For most of humanity, these are things that matter and which give them a sense of well-being, purpose, and what they like to call spirituality. Science is also seen as removing a sense of mystery from life, of rendering life itself into a molecular collection of structures and processes that can be manipulated and used in frightening ways. Science takes its lumps from science fiction views of the future. It gives us models of Faust and Frankenstein as scientists who have lost their moral compass or their empathy for others.

Scientists are often puzzled by the public's understanding of and reaction to science. They do not understand why superstition, myth, pseudoscience, and the paranormal are so popular and widely accepted as legitimate science. They blame public ignorance and talk of scientific illiteracy with the vain hope that somehow a good course in science will dispel these alternate ways of seeing the universe. They also differ on where that scientific literacy should begin and how much of a steeping in real science is needed. Those who are not scientists often react to these proposals as attempts at indoctrination. Some scientists actually accept the attempt to popularize science as indoctrination but favor it as a political move to keep science from being destroyed by Luddites or impoverished by withdrawal of government funding.

To compound the difficulty that science faces, there are many scholars and scientists who disagree over what science is and ought to be. Some critics deny that science tells us anything more than philosophy or religious scriptures. They accuse scientists of pretending to be objective, laying claim to an underlying reality that doesn't exist, and engaging in political behavior that betrays their real intentions as scientists.

I am a scientist and have lived a full professional life as a scientist for the past fifty or so years. I love what I do and I have long reflected on science and scientists. My purpose for writing this book is simple. The public needs some idea of what science is in its many characteristics, including its weaknesses and distinguishing features. I am puzzled by the abstract and remote style that represents much of the writing on the philosophy and sociology of science, which is too removed from the working world of the scientists for them to recognize it as familiar.

It is my hope that those who read this book will have a better understanding of science. I discuss how knowledge is attained through many different approaches and I raise a number of issues that the critics of science raise. It is not my intent to propagandize for science or to engage in polemic debates in defense of science. I think the critics have legitimate concerns and I want them to see how a scientist, in a non-belligerent way, looks at these issues. I am sure I don't represent the views of many scientists because the field of science is too diverse to have a common set of beliefs. I hope too that young people who intend to enter science will see it as a fully human activity and appreciate that they will also have to confront their weaknesses and values when they enter their professions.

My views on science are based on the experiences and choices I have had in my professional career. I have been a geneticist and six students have received their PhDs under my mentoring. I am also a historian of science and I have had 15 books published on science. I have taught undergraduate and graduate courses for biology majors, non-science majors, and medical students. I have attended and participated in numerous national and international programs and I have reached out to the public with a newspaper column, *Life Lines*, for *North Shore Long Island* newspapers since 1997.

My interest in writing a book on the nature and significance of science stems from my 17-year association with the Workshops on the Liberal Arts held at Colorado College in Colorado Springs. Laura Bornholdt (1913–2012) conceived of this program first for the Danforth Association and later for the Lilly Endowment, which funded these remarkable opportunities for scholars and administrators from about 20 different universities and colleges each year to get together for two weeks each summer to discuss problems in higher education and respond to the seminars offered on the liberal arts, including science. At the first of these meetings, I had the good fortune of meeting Sheila Tobias and learning from her about math anxiety in females (although many males also have this fear) and how she found ways for them to overcome those fears.

Much of my outlook on science was shaped by my being mentored by H. J. Muller (1890–1967) at Indiana University in 1953–1958 for my PhD. Muller thought deeply about science and society and frequently found himself criticized and threatened for his views against racism, sexism, and spurious elitism. He raised a concern for protecting the public from radiation exposure. He introduced the concept of genetic load in the population. He offered a program of germinal choice as a compensation for the "load of mutations", as he called it, that civilization imposes on our species.

Besides Muller, I appreciate the conversations I have had about science with Bentley Glass, C. N. Yang, and Linus Pauling on several occasions. I appreciate similar conversations with Barbara McClintock and Jim Watson. I have also benefited from the numerous discussions with my colleagues, especially Gleb Krotkov at Queen's University, J. Richard Whittaker at UCLA, and Abraham Krikorian, Paul Adams, and Paul Bingham at Stony Brook University.

Elof Axel Carlson

# Contents

# 1 Science as Exploration

Most of what we know about the universe comes from exploration. Scientists attempt to describe and go to places looking for interesting things to see. In some fields of science, this is their most important and time-consuming activity. It is not an easy task to describe the known universe. How many species are there? How many stars are in our galaxy? How many galaxies are in the known universe? How many minerals are present in the earth's crust? Some of this sounds tedious and uninteresting: How many bones are in the human body? How many muscles? What are the major blood vessels and nerves? And some of it sounds so trivial that perhaps any of us could figure it out: How many teeth are normally present in our mouths? It is not as easy to find out how many cells are present in an adult body. Some of my colleagues disparage this activity as "stamp-collecting" and consider it unworthy of an intellect's mind. Yet without that inventory of information about everything, we would have a tough time doing science. Data are the food that feeds the scientific mind.

## Carl Linnaeus and the Lapland Voyages

One of the great scientists whose reputation rests solidly with exploration is Carl Linnaeus (1707–1778), the Swedish botanist who founded modern classification. Linnaeus was talented, ambitious, hardworking (driven would be an appropriate term), and egotistical. He was a well-connected minister's son and quickly stood out as an avid collector with a gift for accurate description and gardening that delighted his wealthy benefactors. At the age of 25, with the backing of the Scientific Society of Uppsala, he went to

Lapland to explore the people, their customs, and the plants and animals that dwelled there. He was astounded by what he saw. Sweden is mostly populated in its south and sparsely populated in the north. The climate is mild near the straits that separate Sweden from Denmark, but in the far north where the Lapps live, it is a frigid arctic. Linnaeus quickly organized a second expedition to explore the middle of Sweden (Darlana), a mountainous area rich in minerals. He was already well acquainted with the natural resources of southern Sweden. His voyages gave him a harvest of new plants to describe and a puzzle to solve. How was he to name these new forms of life? How should they be organized or classified?

Although an insult is intended when one scientist calls another a stamp collector, I actually see it as a compliment. Stamp collectors are very organized people. They patiently collect, mount, and display in some chronological or topical way hundreds or thousands of stamps. Collection without organization is not very informative, but good organization reveals a lot. Stamp collections might teach the history of a country and its biases, cultural icons, economy as the prices fluctuate (and stamps are sometimes struck over with new values), aesthetic standards, and pandering to public tastes. Scientists often require an immersion in data before they can see how things are connected.

Linnaeus was 28 when he proposed a system of classification that worked and which he continued to modify until his death, always seeking to classify all animals and plants that transcended local languages and popular names. He sent his students on expeditions and received plants from all over the world. He organized plants and animals into a hierarchy of categories based on some recurring and prominent features. For flowering plants, it was the wispy structures we call stamens and pistils, or the sexual organs of plants. It was a scandalous method in an era that had strong inhibitions about sexual matters. Many of Linnaeus' initial criteria, including his sexual system, have been replaced, but his simple use of a two-name system to classify organisms prevails. We are *Homo sapiens,* which comprise of two Latin words for "man the wise", the term *Homo* being Latin for human, and both *Homo* and human share origins from the Latin term for earth, the humus, that made us and inters us. *Homo* is our generic name and we belong to the species called *sapiens.* We were the only known species of humans during Linnaeus' time. With the discovery of fossil bones of our ancestors, we now recognize other long extinct species, including *Homo erectus* and *Homo habilis*. This dual naming (binomial nomenclature) was Linnaeus' great contribution to simplifying the way all newly discovered species could be uniquely named.

## Charles De La Condamine Explores the Amazon

A contemporary of Linnaeus, Charles de la Condamine (1701–1774), was a mathematician who was active in French scientific societies while they were engaged in a debate between supporters of Isaac Newton and supporters of their own astronomer, Jacques Cassini (1677–1776). Newton had argued that the earth had a longer diameter across its equator than across its poles. He attributed this flattening of the earth at the poles to the effects of gravitation. Cassini argued the reverse and believed that the spin of the earth made it elongate as it spread out its mass along its north-south axis. La Condamine was sent to Peru to measure the meridian northward until he came to the equator while another team was sent to Lapland. The measurements in Equador were taken under difficult conditions, both physical and political. A revolution broke out and La Condamine was cut off from sailing back to France, so he decided to sail down the Amazon starting from its source in the Andes. No one had done this before and there were no maps to guide him. The exploration was thorough. He visited tribes that Europeans had never seen before, described the fish and other animals that lived in the Amazon, and collected everything he could stash in boxes. It took him ten years to get to the mouth of the Amazon in Brazil and he arrived back in Paris with an Amazonian wife, a detailed map of the river he had explored, a log book of his observations, and a sampling of his explorations (most of it was lost through misadventures). Much to his credit, he confirmed Newton's prediction, to the disappointment of his own countrymen who supported Cassini.

## Alexander Humboldt and *The Cosmos*

Alexander Humboldt (1769–1859) inspired generations of scientists with his explorations. He began publishing local explorations of the plants around the mines of Freiberg and made a tour of Swiss and Italian wildernesses looking for new plants. At the age of 30, he was on his way to South America and quickly explored the Orinoco River in Venezuela and traced its connections to the Amazon. He made his way across the Andes and climbed Mount Chimborazo, describing the changes in plant life as the elevation increased. His gift for describing what he saw excited European readers, and his bravery in climbing an unexplored mountain at great risk to his own safety made him a public hero. He became a pioneer founder in the fields of geography, ecology, and meteorology, fields heavily dependent on observation and accurate notes. Humboldt explored every aspect of the universe he could. He measured the transit of Mercury, observed and calculated the periodicity of meteor showers, studied electric currents in nerves, introduced guano (bird dung accumulated over thousands of years) as fertilizer into European farms, measured the earth's magnetic force, plotted the first isotherms for weather mapping, correctly interpreted the mechanism of volcano formation, and reclassified minerals based on his knowledge of their formation. Over a twenty-year stay in Paris after his return to Europe, he wrote his observations and reflections on his readings about nature and the universe. *Cosmos*, the title he chose for this work, appeared in five volumes beginning in 1845.

Humboldt lived to be 90. Much of his life was tied up in diplomatic missions for Europe's kings who trusted him. Humboldt was the first to create a worldwide science network for the study of the earth's magnetic fields and for the study of weather patterns. Wherever Humboldt traveled, he was given a generous welcome because he had no desire to exploit and no political advantage to promote.

## Charles Darwin and the *Voyage of the Beagle*

Charles Darwin (1809–1882) was well-aware of the accomplishments of Humboldt and showed equal zeal as an explorer and collector. However, his father did not intend for him to pursue such a life and wanted Darwin to be a physician like him and his grandfather. Darwin gave medicine a try but quickly gave up on medical school after one year. His father then steered him into the ministry, assuming that a life as a curate in the country was safer than that of a wealthy squire wasting time chasing and shooting foxes and ducks as an idle gentleman. Darwin came from wealth. While in Cambridge,

Darwin discovered to his pleasure that naturalists found a home teaching in seminaries. This was still the era of "natural theology," the belief that the wilderness was a "bible of nature" waiting to be explored, classified, and interpreted so that God's designs and handiwork would be revealed. The chief spokesman for natural theology was the gentle, happy, and kindly naturalist-theologian, William Paley (1743–1805). Paley argued that a human life was far more complex and filled with "mechanical contrivances" than the most sophisticated watch constructed by the finest minds. Yet if we found such a watch while walking in a heath, Paley assured us, we would know instantly that it is a creation designed and put together by our fellow human beings. Encountering a rock on the heath would suggest it was always there or its presence indeterminate but certainly not the handiwork of a human, like the watch. No such thought of a rock-like status would enter the minds of Paley's readers when contemplating any living thing, which are too complex to have been made by chance, by ourselves, or by any intelligence other than the Creator — God himself. It is a powerful argument, and I still hear it being used by my students today when they challenge the theory of evolution by natural selection. As a college student, Darwin had no quarrel with Paley and he fully agreed with these arguments. The young Darwin had distinguished himself as a keen explorer and observer when he accompanied his teachers on their walks in the heaths and along the shores of the less frequented regions of Great Britain. At this stage of his life, the young Darwin was far from being an evolutionist. He believed in the principles of the church of England, including the belief that all life on earth was created, essentially as it looks now, during the six days of Creation described in Genesis.

Darwin's theory of evolution through natural selection would not have been possible without his participation in a voyage around the world that came his way because his own professor, who was scheduled to take that voyage, had just married and did not wish to be separated from his bride. It was also Darwin's good fortune that his Uncle Josiah Wedgwood (1703–1805), of the porcelain dinnerware fortune, persuaded Darwin's father to let his son go. Thus began a five-year adventure that is unparalleled in the amount of observation and exploration leading to a major scientific theory. Darwin left England on the HMS Beagle in 1831, a virtually unknown naturalist planning a career as a minister in the church of England, and returned in 1836, recognized by the best naturalists in England as one of their own, and recognized by himself as a deeply troubled young man who no longer saw the wilderness as a bible of nature and suspected that species today differ profoundly from those at the time of the creation of life. In rejecting natural theology, Darwin realized that he was an incipient heretic who had to be cautious about publicly expressing his doubts.

Darwin wrote well and his attention to detail was remarkable. The *Voyage of the Beagle*, his first book, expressed no doubts about the prevailing beliefs that species were created by God and that they are fixed in their characteristics. What the book does is much more subtle and revolutionary. Darwin astounded readers with one example after another of the adaptations of animals and plants to their circumstances. We see this when he shifted from organisms in Tenerife to those in Brazil or Argentina or the Falkland Islands. He showed us the sloths in the pampas and revealed the fossils of extinct giant sloths found in the same country. Almost every page of his diary took my breath away: "On another occasion I started early and walked to the Gavia, or topsail mountain. The air was delightfully cool and fragrant; and the drops of dew still glittered on the leaves of the large lilaceous plants, which shaded the streamlets of clear water. Sitting down on a block of granite, it was delightful to watch the various insects and birds as they flew past. The humming-birds seem particularly fond of such shady retired spots. Whenever I saw these little creatures buzzing round a flower, with their wings vibrating so rapidly as to be scarcely visible, I was reminded of the sphinx moths: their movements and habits are indeed, in many respects similar." [June 27, 1832 Rio de Janeiro in *The Beagle Record*, edited by R. D. Keynes]. This is sort of the beguiling writing of a country parson, an Isaac Walton extolling fishing.

But a few paragraphs later, one reads: "I was much surprised at the habits of *Papilio feronia*. This butterfly is not uncommon, and generally frequents the orange groves. Although a high flier, yet it very frequently alights on the trunks of trees. On these occasions its head is invariably placed downwards; and its wings are expanded in the horizontal plane, instead of being folded vertically, as is commonly the case. This is the only butterfly which I have ever seen that uses its legs for running. Not being aware of this fact, the insect, more than once, as I cautiously approached with my forceps, shuffled on one side just as the instrument was on the point of closing, and thus escaped. But a far more singular fact, is the power which this species possesses of making a noise. Several times when a pair, probably male and female, were chasing each other in an irregular course, they passed within a few yards of me; and I distinctly heard a clicking noise, similar to that produced by a toothed wheel passing under a spring catch. The noise was continued at short intervals, and could be distinguished at about twenty yards distance. I cannot form a conjecture how it is produced; but I am certain there is no error in the observation." [p. 65] This was the emergent Darwin, the naturalist who was so transfixed by nature that he drew every minute fact he could out of an observation. He was agile, alert, curious, and enthusiastic. Where we see a butterfly resting on a tree, he revealed a habit of avoidance that is atypical and thus advantageous to the survival of the butterfly. He doesn't tell us that, but in retrospect, we can see the curiosity churning in his mind. Was he gathering this information to amuse us or to stimulate his own imagination and to find explanations for this unusual behavior? Great theories take time to emerge from masses of data and Darwin had no such theory in 1832. It would take another six years before the idea took root in his mind that the thousands of adaptive responses he observed had a common significance. I suspect that he sought a reconciliation with natural theology while sailing home on the *Beagle*. If God is the great designer of life that William Paley had extolled in his writings and that Darwin had absorbed as a student at Cambridge, why would the "bible of nature" be so contradictory to what theologians thought God did? Why would the species in Tenerife off the coast of northwest Africa resemble African species and not the species of the Galapagos, off the northwest coast of South America, which in fact resemble South American species? The islands are both volcanic, similar in climate and their distance from the mainland, yet their two sets of organisms were profoundly different from one another. When ideas based on religious faith make no sense and there is an abundant mass of data that begins to make sense, a scientist goes with the data. If the "bible of nature" leads not to some theologian's interpretation of *Genesis* but to evolution by natural selection, then heresy is born. Darwin knew this and, for some 20 years, kept these thoughts to himself and shared them only with his closest scientific friends.

# Darwin's Evolutionary Theories Led to Controversy

When *The Origin of Species by Natural Selection* first appeared in 1859, Darwin was well prepared to make a solid argument. He had devoted years to visiting breeders at agricultural, horticultural, and pet shows, corresponded with hundreds of naturalists, and read extensively of the curiosities observed by others as the British Empire extended its way around the world. He had originally planned something like a ten-volume approach, outdoing Humboldt's *Cosmos* in its completeness, but the pressure of Alfred Henry Wallace's (1823–1913) independent interpretation based on far less evidence galvanized Darwin into writing a short abstract of those volumes in progress. Darwin presented a short paper with Wallace's, both being read at the Linnaean Society to the scientific public the year before. When Darwin's book appeared, the entire first edition was sold out on its first day of issue.

Darwin is still controversial to this day, but less among scientists than among those who cannot surrender their interpretation of the creation of life based on religious scripture. Although the major religions in America and Europe have accepted evolution as a reality and the theory of natural selection as a plausible and legitimate scientific explanation of that evolution, the majority of people in these countries are still doubtful that life had an evolutionary origin. When we contrast Darwin's explorations with those of Linnaeus, la Condamine, and Humboldt, we are struck by the difference in reception. Until 1858, Darwin belonged to these celebrated explorers who advanced our knowledge of the universe and satisfied our curiosity about what is out there by elucidating the vague mechanisms of how things worked, from a ship's magnet to the storms predicted by weather fronts far from the site of their future damage. People enjoyed the order provided by Linnaeus's taxonomy and celebrated the courage of these men who faced high risks in a world of strange diseases, potentially hostile people, and uncharted territories. For Darwin the evolutionist, the public reception was largely disapproval for how he had abandoned God, corrupted their children's education, and robbed them of their favored status in the universe. It is one thing for God to remind us that we came out of dust and will return to dust, but it is unseemly when we learn that the earth did that and does that regardless of what we believe!

## Our Knowledge of Space and its Components Relies on Exploration

In our own time, much of science is still done by exploration. I was thrilled by the landings on the moon and Mars, the radar probes of Venus, and the close-up video pictures of the outer planets and their moons. Without astronauts landing on the moon and bringing back a few sacks of rocks, we would not have direct evidence of the age of the moon and how its mineral composition differs from that of the earth's. Someday we will have rocks from Mars, not heaved at us from meteoric impacts but from our own unmanned or manned space voyages.

Before those explorations, there was only speculation to guide us. Did the moon have a thin layer of dust on its ground or was it several feet thick? If an explosive charge is made on the surface of the moon, what seismic response will it produce? Is it a solid rock or does it have a liquid center like the earth? What is the composition of the rings of Saturn? What is the "eye" in Jupiter's middle region? Why are there belts of different color on Jupiter and what do they imply? Are the moons of Jupiter and Saturn just like our own — a pockmarked, desolate, and arid wasteland without an atmosphere?

We now know answers to many of these questions from space exploration. For astronomers, the Hubble space telescope, launched with great hope, was at first ridiculed for its myopic and out-of phase-lens and restored to its full potential with some corrective optical engineering. The resolution of its images is startling, and the Hubble continues to astound astronomers with pictures of a larger and more complex universe, making its interpretation all the more difficult for astronomers. I have followed the cosmology story since my high school days and first took turns reading George Gamow's big bang model and Fred Hoyle's steady state model, after which came Hannes Alfven's models of galaxy-sized cosmic plasmas, recycling the universe in unending perpetuity. I have no idea how the arguments will play out, but on one thing all scientists agree — the more data astronomers have, the more likely their models will reflect the history of our universe. Data is amply provided by the many space launches to observe galaxy formation, black holes, the death of stars, the terrain of Mars, and the atmosphere of outer planets.

Astronomy is too remote for most of us on earth. We see only a smattering of stars and most of humanity has never looked through a telescope. I bought a telescope for my children one Christmas some forty years ago and I recall the thrill I felt as I saw Jupiter's moons. Like Galileo, we plotted their positions for several days and indeed saw that they orbited that planet. As I was using a far better optical system than what was available to Galileo (my Saturn showed rings, his Saturn showed "ears"), I admired that he could extract so much more out of his observations than I could. What distinguishes him from most amateurs is more than accuracy of description. He could use it, reflect on it, and see connections that most of us lack the talent to develop.

## Exploration was the Basis for Anatomy

All of us, however, have some familiarity with our own bodies. I am quite surprised at how little of it most students know. It is not until they take a biology course and dissect animals that they discover what their insides are like, but even this is a modest inventory of major organs and major blood vessels. It is not until a medical student takes gross anatomy that the overwhelming amount of things in the body becomes apparent. There are names for everything and functions for them as well. Each name and each function had to be worked out by some adventurous explorer who described a part of our anatomy no one previously had looked at. Each function required lots of work to interpret. If it is a muscle, how does it work? If it is a gland, what does it secrete? There had to be a first person to recognize that organs were composed of tissues (his name was Francois Xavier Bichat), that tissues were composed of cells (his name was Matthew Schleiden), that cells contained nuclei (his name was Robert Brown), that nuclei were composed largely of nucleic acid (his name was Freidrich Meischer), that the nucleus had its contents organized mostly as chromosomes (his name was Henirich Waldeyer), that the chromosomes consisted of sequences of genes (his name was Thomas Hunt Morgan), and that genes were organized as double helixes of DNA (their names were James Watson and Francis Crick). Did you also note the shift from single authors of books and scientific articles for most of the history of science until about the middle of the twentieth century? Did you notice the absence of women scientists and how long it took before women could freely enter science in Europe and North America where these events took place?

The twenty-first century's catalog of human parts will not be *Gray's Anatomy*; it will be a multi-volume listing (probably as an e-book on a computer) of the 23,000 or so genes that constitute the human genome. The human genome project is rapidly mapping all of our genes, sequencing each gene to its individual nucleotides, determining the function of a substantial number of those genes from already known sequences that reveal its function, and identifying when and where that gene will be turned on or off in our life cycle. It will be the ultimate *Scott Catalog* to satisfy the stamp collector in all of us.

# 2 Science as Discovery

"Treasure your exceptions." This was the advice of William Bateson (1861–1926), the British geneticist who named the field of genetics and battled with his colleagues to get Mendel's findings established as the basis for modern heredity. Many a scientist will endorse those words. When I was a graduate student, I remember finding an unusual fly that ought not to have been there. I had been doing an exercise for a graduate genetics laboratory course that Nobelist Herman Muller (1890–1967) taught every year at Indiana University. I called the teaching assistant over and he swept it into a vial for Muller to look at. Several months later, Muller plunked a one-page abstract on my desk. I saw my name as a co-author and I protested that I didn't deserve to have my name on the paper since I did no work for it. "If you hadn't found that fly, there would be no paper," he cheerfully replied and left the room before I could protest further. That fly and that paper became my dissertation research.

Discovery involves both luck and readiness. Critics of science often claim that scientists have their ideas before they gather their data or conduct their experiments. If that is so, many a good experiment or finding would have been discarded because the discoverer did not treasure an exception. I remember one such incident in Muller's laboratory a few years later. Helen Meyer, one of Muller's associates, noted that the mutation rate in one batch of fruit flies she had analyzed was consistently higher than another batch that was analyzed later that same day with the same ultraviolet light exposure. The only explanation she could come up with was that she had stacked the boxes containing the exposed fruit fly eggs one on top of the other and the bottom layer gave rise to the higher mutation rate. She never pursued the mystery further, but a year or two later, she read of a new phenomenon called photo-reactivation whereby exposure to visible light after ultraviolet radiation treatment prevented much of the damage to DNA that ultraviolet normally causes. We know too well what ultraviolet radiation in large doses can do to our skin from early summer sunburns to later onsets of skin cancers and melanomas.

## Classical Genetics was Dependent on Discovering New Mutations

Discovery is one of the most pleasurable experiences in a scientist's career. To find something that no one else has recognized before and to be the first to see, describe, or interpret it is exhilarating. In the early days of the fruit fly laboratory at Columbia University, Thomas Hunt Morgan (1866–1945) and his students found new mutations at irregular intervals. There were 15 found in 1910 and all were by Morgan. Then there were 10 more in 1911, 24 in 1912, 31 in 1913, and 16 in 1914, and most of those mutations were found by Morgan's students. The basis of classical genetics was established during this remarkable period in Morgan's laboratory — X-linked inheritance was found and interpreted, crossing over was discovered and applied to the mapping of genes on their chromosomes, the theory of the gene was established, non-disjunction was discovered, and at least one gene had been found for each of the four pairs of chromosomes that composed fruit flies. Muller, who was intrigued by mutation rates, told us that when any of the students found a mutation, the laboratory celebrated with a bottle of wine. In the 1920s, Muller described mutations as being so rare that it was like finding a dollar bill while walking the streets of New York; you did not think of frequency and just considered yourself lucky to have found one.

Discovery comes in many forms. I can imagine the surprise Roentgen must have felt in 1895 when he set an electric current through a glass tube and noticed a nearby panel beginning to fluoresce. His curiosity aroused, he began experimenting and in one of his memorable demonstrations showed a photograph of his hand with its gold ring like a fallen halo around his ring finger, the bones nicely revealed and the flesh a faint ectoplasm around them. When paleontologists look for hominid fossils, they are well prepared to find dozens of broken bones and a few immediately recognizable teeth that signify the presence of a primate skull, probably human. It may take a year or two to clean those bones of their matrix of rock and assemble them into a skull. It will also take time to date the rock and to identify the layer in which it was found as well as the other debris, especially other mammal fossil bones, in that layer to reveal the circumstances associated with that fossil, such as the animals that resided there, whether they served as food, and how they were killed and eaten. However, the image of archaeologists and paleontologists presented in the movies is often very different and we incorrectly see them as treasure hunters who may or may not be lucky in making a find. The effort is often deliberate and planned, with the location chosen for a variety of suggestive reasons. The true surprise is the unexpected finding.

When Muller applied x-rays to male fruit flies in the fall of 1926 and looked for a category of mutations using a special genetic stock he had designed, he found these mutations not with the rarity of his student days before World War I but instead dozens each night. He could hardly contain his excitement and every few minutes opened his window and shouted to a colleague in a laboratory beneath him, "Hey, Bucholz, I found another!" Here, the discovery was not the novelty of something new but the unexpected frequency. Muller had assumed for years that x-rays would only double or triple the spontaneous frequency (rare as those dollar bills were on the streets of New York!); he did not expect the rate to zoom upward to 150 times the spontaneous frequency.

## Discoveries Lead to Theories

A discovery may not jump straight to the eye but instead take a lot of sleuthing to figure out. Imagine young Alfred Sturtevant (1891–1970), a sophomore in Morgan's class. Sturtevant came to Columbia from Alabama where his father had a large farm. His brother Edgar was a professor of philology (the study of language evolution) at Barnard College, and young Sturtevant roomed with his family. Morgan was teaching introductory biology only because the regular instructor was on sabbatical, and Sturtevant was taking the course only because his brother told him to. Edgar knew of his brother's interest in horses back on the farm and that he had gone to the N. Y. Public Library to learn about their pedigrees and how horses inherited their coat colors. Sturtevant's interest in color may have also stemmed from his own red-green color blindness; he sometimes competed with his brothers in strawberry picking and could never gather as many strawberries among the green leaves as they! Sturtevant's interest in color pursued him to his old age. When I visited him at Pasadena in 1966, he took me to his iris garden and showed with pride the wonderful hybrids he had created and which he thought were so beautiful. I did not have the heart to tell him that they were the least attractive irises I had ever seen! He had created a small field of puce, mustard, and faded mauve irises.

Morgan, who was impressed by Sturtevant's efforts to study the genetics of thoroughbreds, had mentioned in class that he had observed something new — while working with fruit flies, he found that genes farther apart on the chromosome produced more crossovers or recombinations than genes that were close to one another on that chromosome. Sturtevant asked Morgan if he could borrow his data and Morgan gave him a copy of the results. The 19 year-old Sturtevant then stayed up all night and constructed the first ever genetic map from Morgan's data! As Sturtevant stared at his map with the location of those five or six genes Morgan had found, I can almost feel his excitement as he waited to share it with Morgan. I suspect that Sturtevant had no clue how to use the data when he took it home, but the idea of mapping the genes based on the data emerged from Morgan's suggestive words on the relationship between distance and frequency of crossing over. Morgan had missed it — he had already published his first few papers on crossing over and was more interested in debunking Bateson's erroneous models than in making a chromosome map, so that idea did not occur to him. It is not rare for very bright scientists (Morgan received the Nobel Prize in 1934) to miss the obvious in the very field they dominate. It is also not rare for someone younger and not saturated in current issues to see things in an uncluttered way.

# Francis Bacon Stressed Induction as Essential for Science

Discovery also comes out of immersing oneself in data. Francis Bacon recognized this some four hundred years ago and called the process "induction". Many skeptics in philosophy and the history of science do not believe that induction occurs. Instead, they believe that scientists have preconceived ideas and use the data to confirm their theories. I have no doubt that much of law works that way — it is adversarial, and a good lawyer with something to prove will seek all the evidence available to bolster that point of view. If I were a client, I would want my lawyer to be my advocate, not a scientist seeking the truth or objectivity of a situation. But most scientists are not lawyers. They want to satisfy their curiosity and not prove their pet hypotheses. With some exceptions, they are glad to dump an idea that is wrong and try again. I know of no scientist who has not experienced failure. Most scientists run pilot experiments before conducting them on a larger scale. They know that their expectations may be very different from what the data will show.

Gregor Mendel (1822–1884) was more likely than not a Baconian scientist. He had as difficult a life as scientists go. He grew up as a peasant's son but lucked out when his village instituted mandatory public education for elementary school classes. He turned out be an excellent student and his teachers encouraged his family to give him more education. His sister sacrificed most of her dowry so that Mendel could go to the equivalent of high school. Once again, he proved his aptitude in class and teachers urged him to further his studies. However, his father injured his back and could not take on more work to help his son. Mendel eventually chose to become a priest in a teaching order. His religious faith was sincere. He was sent to the University of Vienna where he took classes with the physicist Christian Doppler (1803–1853) and one of Europe's finest mathematicians, Andreas von Ettinghausen, who introduced him to combinatorial algebra, as well as with one of Schleiden's students who taught him about cell theory. Mendel hoped to become a physics teacher and he tried twice for a license but flunked the examinations both times. He returned without a degree and remained a substitute science teacher in his Moravian monastery until he was elected to be the abbot of his fellow monks. He had no stomach for the priesthood and felt faint at the sight of ill people.

## Mendel Inferred his Laws of Transmission From Years of Growing Peas

Shortly after his return from Vienna, Mendel began to breed flowers. He liked horticulture and frequently grew plants on his father's farm. He used the monastery garden and chose peas as his subject of choice. After obtaining different varieties of peas from several seed suppliers, Mendel bred them to see how true they were to their described type and what would happen if he hybridized them. No doubt an incipient stamp collector, Mendel kept careful records of each cross and the progeny peas that were gathered from the pods. The records accumulated each year. If the traits he chose were not stable and proved too difficult to maintain, he discarded them and focused on the "good" traits. He also chose traits that did not blend, such as red versus white flowers and no pink flowers, hairy leaves versus smooth leaves and none with irregular stubble, and yellow versus green peas and none that were chartreuse. He figured out how to pollinate the plants and keep them from accidental pollination by other insects, such as the pea weevil which ruined many of his plants. It took him seven years of data gathering before he presented his results. We do not know when the idea of hereditary units came to him, when he saw that some traits were recessive (they did not manifest in the hybrid) and others were dominant (they looked the same in the hybrid as they did from the pure stock used by the seed supplier), when he conceived of statistically making ratios of the traits among the different generations of offspring, or when he conceived of using combinatorial algebra to express his general laws. There is one simple reason for our ignorance. Later in life, Mendel went against church policy and defended his monastery against taxation from the government, much to the annoyance of his bishop and the monarchy who were willing to settle the tax bill for a pittance. Mendel was not very tactful in his letters to bureaucrats. On one occasion, he locked himself in his office, refusing to let in tax collectors, and he was carried out in his chair by officers of the Crown. When Mendel died of kidney disease, his monks burned all his correspondence and notes, including the scientific notes for his experiments.

Mendel published two articles, and his first article concerned the laws of heredity. If we are to believe Mendel, he inferred them, thus essentially doing what Francis Bacon said scientists should do. He saw patterns or connections among the data and expressed them mathematically. In the 1860s, most botanists and zoologists had little to do with mathematics. Some who had heard of or read his paper thought that this was a form of numerology or some cabalistic playing with numbers. Mendel sent out some 40 copies of his article (scientists call these reprints in the US and offprints in Europe) and corresponded with Carl Nägeli (1817–1891), another of Schleiden's students and one of the leaders in studies of heredity at that time. He sent packets of peas to Nägeli and predicted what the ratios would be and how he obtained these hybrid seeds. As Nägeli was a scholar, he knew what was going on from his vast readings and contacts with those in the field of hybridization. He told Mendel that the material he was working with, *Hieracium* or also known as the hawkweed, did not Mendelize, so his laws were not universal. He sent Mendel some hawkweed seeds, and Mendel grew them and crossed them and sure enough, Nägeli was right — their heredity was very different. Mendel's second paper is essentially a retraction of his first paper. He could not explain why hawkweeds differed and for all he knew, peas were the exception, not the rule. Mendel began to believe this when he switched to bees and they were even more bizarre in their heredity than hawkweeds. He gave up on heredity studies and switched to meteorology, possibly inspired by Humboldt's success in establishing weather stations and weather reporting throughout Europe.

Hawkweeds did not yield their secrets for another fifty years. It turned out that its pollen does not penetrate the ovule and its nuclei do not fertilize the egg nucleus and the endoderm nucleus of the embryo sac. Instead, the pollen tube reaches the membrane of the embryo sac and stimulates the egg nucleus to begin dividing and forming a maternal embryo. In about 2% of cases, an actual fertilization takes place. Thus, the results are messy and look mostly like maternal inheritance. In the biological realm, diversity rules supreme and no two species are alike. Indeed, even within a species, no two individuals are alike. This diversity creates complications rarely encountered by chemists and physicists, who work with identical atoms or molecules that number about ten to the twenty-third power, a number so huge that it bankrupts our imagination even though we stir that many molecules of sugar in a cup of tea or coffee!

Scientists can discover things like fossil footprints, hominid ancestors, Precambrian fossils, possible life forms on Mars, supernovae, and comets, principles such as X-linked inheritance, meiosis, acceleration, velocity, and chemical bonding, and broad theories such as gravitation, relativity, natural selection, cell theory, and the gene as the basis of life. Sometimes scientists discover something that they or others have predicted; other times they discover something they weren't looking for. Both are important ways that science works.

## We do not Know How Insight or Induction Arises

I am not persuaded that scientists are visited by flashes of revelation and that they make their discoveries through dreams, daydreams, or hallucinations. Some have claimed that they did, although their claims have been disputed. Friedrich Kekulé (1829–1896) was alleged to have dozed before a fire and imagined salamanders chasing their tails, and from this he inferred a circular or ring shape for what he previously thought was a sequence of six carbon atoms. The benzene ring, Kekulé's discovery, is thus regularly taught to chemistry students. The dream is suspicious because it makes use of an older legend buried in mythology — the ourobourus or snake that consumes its own tail, a symbol used by mystic sects in the years when Christianity was competing with other imported religions in the Roman Empire. How much is legend we don't know, but many scientific discoveries are associated with such legends, from apples falling on Newton's head to Galileo figuring out the laws of falling bodies by dropping large and small iron balls from the tower of Pisa.

Muller did mention in one class I attended that Morgan conceived of crossing over when staring at a railroad timetable showing blocks of scheduled times for AM and **PM** in Roman and **Bold** fonts, the latter having a darker and thicker appearance. It suggested to him that blocks of genes were exchanged rather than individual genes hopping from one chromosome to the other. I don't know if Morgan told this to his class as a means of illustrating his theory or as the explanation for his insight, but often what is said in class can be turned around from one intention to another.

## Some Scientists Attributed Insight to their Dreams

I wish I had a personal experience in which a moment of discovery came as a dream or hallucination. I am curiously devoid of such experiences. I can think of two instances, one mystical and the other scientific, that approximate Kekule's alleged vision. As a youth, I enjoyed reading different works of religion because I was raised an atheist and did not experience the rites of passage and weekly activities that my religious friends enjoyed. I could thus read religious writings dispassionately and I enjoyed them because they satisfied my curiosity about the world of my classmates. I had read St. Francis's *Little Flowers*, an account of his mystic experiences as a medieval monk, and the biography of Sri Ramakrishna, a nineteenth century Hindu holy man. Both shared some very vivid mystical experiences in their encounters with their followers and friends. During the summer, I worked as an elevator operator at 217 Broadway next to St. Paul's Chapel in the New York City Hall district. In the early 1950s, elevators were hand controlled and there was a person, called a starter, who directed the movement of elevators to prevent the cars from being all up or all down at the same time. He instructed us to move our car just beyond the top floor and rest there with the lights out, waiting until the light of the neighboring car was below us before beginning our descent. Sometimes three or four minutes would elapse and for a restless person like me, this was a waste of time. On one occasion, I reflected on the states of satori that Ramakrishna experienced and how he would merge into oneness with the divine. My fingers on the brass gate seemed to merge with it and, for a brief moment, I felt as if I were undergoing the trance-like state of satori. The light of the upcoming car, however, cast a bit of reality on my reverie and I was quickly on my way down, back in the material reality that my passengers, I am sure, preferred me to experience.

A similar rush of euphoria hit me when I was working late one night constructing a genetic stock for my dissertation research. Among the flies was a totally unexpected one. I thought that it might be a contaminant, but when I looked at the genetic markers it exhibited, I realized with a flash that it was caused by a crossover within a pair of related genes I was studying and that a previous model I had been using was wrong. I saw immediately that this opened up a new line of crosses, and a much more exciting micromap of the gene I was studying would emerge. All of this realization must have happened in less than a minute of looking and reflecting on the fly. I wanted to run to the other rooms but my fellow graduate students had all gone home! Here I was with such a joyous experience and no one with whom I could share it.

# 3 Science as Invention

The nature of inference is largely unknown. Why should this surprise us? When we are engaged in conversation, we do not plan each sentence prior to its tumbling out of our mouths; only the tongue-tied would attempt that. When we write, something very similar happens. The words pour out and from time to time, we back away from them and revise. But the brain is so efficient that it can take lots of data and organize it for us without apparent effort on our part to force it. Of course, as every teacher or writer knows, there has to be preparation, knowledge of the topic, and motivation to express ideas in words whether spoken or written. Artists also experience this rush of expression in painting, sculpting, dancing, acting, and composing. Socrates was troubled by this gift of playwrights. They wrote magnificent plays, but when he interviewed them, he found them wanting in clarity, logic, and understanding of their creative process and the source of their inspired phrases. Socrates concluded that a daemon possesses the creative person and that we are merely agents for these inner daemons who do our creative work. We no longer accept this explanation, but we are no wiser today than in Socrates' time in terms of our understanding of the creative process and how words, ideas, and theories are invented.

## Model Building and Science

I shall present four successful inventions that were powerfully influential in their fields and a bit about the circumstances of each invention. By invention I do not mean mechanical tools, which we will discuss in a different chapter, but rather mental inventions. They are tools of the mind used to develop a field of science and are not the same as theories or hypotheses.

The first of these, which is the most familiar, is the use of model building to construct macromolecules. Its history goes back farther than the success of James Watson and Francis Crick in using it to solve the structure of DNA. The immediate inspiration for Watson and Crick was Linus Pauling and his successful application of visual models of protein sheets and folding based on the inferred alignment of amino acid side groups. Watson and Crick took this one step further by using paper and metal cut outs to fit molecules together and constructing their double helix based on x-ray diffraction studies of DNA molecules prepared by Maurice Wilkins and Rosalind Franklin. The failure of those competing with Watson and Crick, I believe, was their reliance on one major approach that Watson and Crick recognized as actually requiring several approaches. For instance, the biochemist Irwin Chargaff relied on classical chemical approaches like the use of digested DNA molecules which revealed certain fixed ratios of nitrogenous bases (adenine, guanine, cytosine, and thymine) in the DNA of different species. All of these ratios were constant and universal except one. If he measured the ratio of adenine plus thymine to that of guanine plus cytosine, the ratio would vary from species to species but remain constant in any tissue of an organism. The constancy among all plants, animals, and microbes of the ratio of adenine plus guanine to thymine plus cytosine and the constancy of the ratio of adenine plus cytosine to guanine plus thymine was expected since the first decade of the twentieth century. In 1914, Phoebus Levene thought that DNA was an uninteresting crystal because of the apparent simplicity in the ratio (he found each of the nucleotides to be equally represented in DNA). This generated a tetranucleotide theory which reduced DNA to the interest level of table salt, which is a boring repetition of chlorine and sodium atoms in a grid. For the next thirty years, biologists and chemists would mistakenly think of DNA as a simple scaffolding on which proteins (presumably genes) were draped.

Chargaff could not infer the three-dimensional structure of DNA with all its essential features from such a test tube approach. No chemist could. It would be similar to the frustration we would experience if we were told to reconstruct a page in a book by reassembling all the isolated words in it. The same frustration would occur from reliance on x-ray diffraction analysis. It is helpful in determining such features as the helicity of DNA and its doubleness, and it tells the distance between each gyre or turn of the DNA molecule. It also can tell if the nitrogenous bases are inside or outside, which is certainly very important and essential. But what it cannot tell is the sequence of bases along the length of the DNA molecule or the complementary association of the two bases that hold the molecule together. It does not reveal the presence of hydrogen bonds that bring about that pairing association.

For this reason, I believe Maurice Wilkins and Rosalind Franklin were bound to lose the race to Watson and Crick, who used model building to supplement the chemical and the x-ray diffraction approaches. Model building allowed Watson and Crick to play with the shapes and angles of the nitrogenous bases in their more complex form as nucleotides and see how they could be made to fit. This was more than just play, however, because Crick had to use his knowledge of physical chemistry to interpret the angles of all the atoms in these molecules, a task that Watson could not do because he lacked the background for it. What I enjoy about Watson's account, *The Double Helix*, is this bit by bit, puzzle-solving approach, which tells a lot to me about how he thinks as a scientist. He is flexible, intensely curious, a hunch player, and willing to connect things that seem to belong to different fields. In Crick he found his intellectual twin, a person who loved to speculate, relate, and try out numerous ideas without feeling foolish if they were wrong. Many scientists are not as inventive and many are very cautious. They may be superb in what they do, but by not going into cognate fields, they may miss out making major contributions to a topic that interests them. Very often, scientists who dip into other fields are looked down on by specialists in one particular field as dilettantes. There is also a tendency for graduate students to focus narrowly and intensely on one small segment of the universe and immerse themselves in the prevailing approaches of that field.

A much more powerful influence on Watson's thinking came from Max Delbrück who served as his ex officio mentor. Delbrück was a founder of the field called molecular biology, which originated in the first quarter of the twentieth century when physicists using x-ray diffraction recognized that they could solve the structures of simple crystals. By the 1930s, some physicists were applying x-ray diffraction analysis to viruses such as the tobacco mosaic virus or to complex molecules such as wool and other animal and plant fibers. Geneticist H. J. Muller was so enthusiastic about these approaches that he wrote an article in 1936 titled *Physics in the attack on the fundamental problems of genetics* and hoped that the genetic material, perhaps as chromosomes, could someday be analyzed that way. Muller's enthusiasm for such approaches stemmed from his own x-ray induction of mutations. Here was a tool of physics being used to "dissect" genes and learn about their pieces and mutated properties. Delbrück was strongly influenced by these attempts of biologists to lean on physics and he tried to reverse the approach by moving into biology. He discounted the value of following traditional biochemical approaches because he was convinced that molecular structure was the key to understanding basic life processes, especially deep questions like "what is life?" or "what is a gene?"

Delbrück thought viruses would be the best system to explore because biologists had considered viruses as "naked genes." Delbrück soon founded, with Salvador Luria and Alfred Hershey, a school for bacteriophage genetics. They worked out the life cycle of bacteriophage, the tools for doing genetic research with bacteriophage, and the first maps of the genes in this organism. It turned out that these viruses were far bigger and more complex than Delbrück had thought them to be. Delbrück's phage school of genetics (pronounced with an *a* as in the word *page*), located chiefly in Cold Spring Harbor in New York and Caltech in Pasadena, differed from the Parisian phage school of genetics (pronounced with an *a* as in the word *wash*), which was more biochemical in its approach. Delbrück made his students and colleagues look at genetics with this fresh approach, which included features such as electron microscopy, x-ray diffraction, target theory, Poisson mathematical analysis, and the use of plaque morphology (holes digested into a carpet of bacteria).

## Every Scientist is Influenced by the Teachings or Habits of Other Scientists

It is no surprise then that the young Watson looked up to Delbrück as his mentor (Watson got his PhD with Luria) because Delbrück had world stature as a physicist who turned to biology and goaded his students to think in original ways and shun conventional approaches to solving problems. When I look at debates about the discovery of the structure of DNA, I am puzzled by this lack of recognition of teaching and intellectual influence that plays such an important role in shaping the scientist. Watson clearly saw the gene as the basis of life because he was convinced of its rightness when he took Muller's graduate classes at Indiana University. He was convinced of Delbrück's approaches because he was part of that molecular school. He seized on Wilkins's x-ray pictures when he attended a seminar in Italy because he had that intuitive flash that this was the way to gain fundamental insights into structure. He devoured Pauling's approach of model building because he recognized that a three-dimensional shape is not evident from chemical analysis. Too many critics focus on competition, rivalry, sexism, conscious or unconscious theft of ideas or data, and aspects of personality that are familiar to all men and women — our frequent pettiness and sexual needs, our occasional immaturity, and our lapses of vanity. These things undoubtedly accompany our lives, but science overrides them. Science is not caused by these quirky universal realities.

Herman Muller is best known for his work in establishing the field of radiation genetics after he successfully demonstrated that x-rays induce mutations. However, his students recognize another feature of his scientific life that is not as fully appreciated. He was the first to produce elaborate genetic stocks to solve genetic problems and design successful genetics experiments. For Mendel, it was sufficient to follow hereditary traits without a sense of where they were. Once Morgan realized that they were on the chromosomes and that they could be mapped (as Sturtevant so beautifully showed), this opened up the possibility of using combinations of genes and complex rearrangements of chromosomes to make genetic stocks. Muller used stock designs to pursue his studies of mutation. He invented a technique for isolating and maintaining forms of mutant genes that kill the embryos. It is one thing to look for white versus red eyes or rudimentary or normal sized wings, but it is not so easy to maintain a culture that harbors a newly arising mutation that kills early in the life cycle. When I interviewed his colleague and lifelong friend, Edgar Altenburg, I learned how Muller introduced his most famous stock design, the *ClB* technique. This stock contained a crossover suppressor, a recessive lethal, and a dominant visible mutation (bar-shaped eyes). Muller and Altenburg had been working in the laboratory when Muller jumped up and told Altenburg that he had constructed a chromosome combination that could be useful for their mutation studies. The stock allowed Muller and anyone he trained to pick up mutations on the X chromosome because they would either produce no sons (if the mutation was an X-linked lethal) or the sons would all show an abnormal feature (if the condition was a recessive visible mutation). The daughters would harbor the mutation but not express it, and these daughters could be mated each generation with appropriate males to maintain that new mutation.

Muller designed stocks to detect complex events such as translocations, female sterile mutations, male sterile mutations, non-disjunctional events, and mutations at specific genes. He used these stocks to measure the rates of production of different mutations or rearrangements with a variety of mutagenic agents. He designed stocks to accumulate mutations over many dozens of generations without losing them. Today, fruit fly geneticists routinely design such stocks for their own experiments, and part of the graduate experience in genetics is mastering the art of stock design. The essential feature of genetic stock design is the introduction of genetic markers, or genes that have nothing to do with what you are interested in but which help to identify the chromosome or event of interest. It also involves the use of rearranged chromosomes with pieces inverted or chromosomes broken and rejoined in unusual ways. These rearrangements are used for tasks such as the prevention of crossing over and thus keeping a combination of genes intact indefinitely. They enable a desired mutation that is lethal, sterile, or too frail as a visible mutation to be kept indefinitely as a stock.

Muller was successful in proving that x-rays induced mutations because he could render the problem of objectivity moot. The category he sought — recessive lethal mutations on the X chromosome — could be counted. Hence, the rate of mutations could be determined for any given dose as well for doses that are administered from one massive release of radiation or administered at exquisitely low doses chronically over a period of 30 days. Fundamental findings that allow us to make important health decisions came from such uses of genetic stocks, which are far superior to what I sometimes call the "science fair approach" to mutation that used to be carried out in my youth. Students in high school would ask a dentist or relative to x-ray fruit flies (what are called wild type or normal flies) and the offspring would be observed for visible mutations. Most students found not mutations but mutilations caused by legs or wings ripped by the gluing effects of food as the fly struggled to free itself. Samples included flies whose fine bristles would be caked with food producing dried out horns or excrescences that would be mistaken as mutant forms of the body. Very frequently, neither the dentist administering radiation nor the student requesting it, knew what was an effective dose to administer to the flies, and I suspect that most fruit fly science fair projects in those days got doses that were well under 100 roentgens, hardly enough to see any real visible mutations that might have been induced as dominant mutations. In most of the experiments that I and other students in Muller's laboratory carried out to study mutations, doses of 1000 to 4000 roentgens were typical. To give an idea of how big that dose is, your chest x-ray is about 0.01 to 0.1 roentgen. A Hiroshima dose that caused severe radiation sickness was about 200 roentgens, and a dose of 400 to 500 roentgens of full body exposure is enough to kill half of the human adults exposed to it.

# A Mathematician's Insight Into Population Genetics

The fourth invention I will discuss was raised in the early days of Mendelian genetics when those laws were being applied to hundreds of plants and animals. William Bateson (1861–1926), an enthusiastic supporter of the new Mendelian genetics, lived in Great Britain, a country that held firmly to its Darwinian theoretical heritage. One of the assumptions in Darwinism is that the major features of an organism change gradually. That is, the variations are subtle and a shift in any one direction is barely seen from generation to generation, but each generation reveals a continuous gradation of character differences. This would be true for traits like height or lifespan. The Darwinians resisted Mendelism because contrasting traits like yellow or green peas were seen as atypical. Nose shapes don't come in two forms, and neither do most of the things that compose an organism. Essentially, both sides misunderstood each other. Bateson spent years writing angry articles to criticize his Darwinian foes for failing to acknowledge discontinuous heredity, and they, in turn, denounced Mendelism as an oddity of no significance to evolution.

   One question that bothered the Darwinians was the appearance of new mutations. Would these soon flood the population of a given species if they were followed across several generations? This question is addressed today by the field of population genetics, which tries to provide mathematical models of the fate of genes in populations. The first effort to solve this question was done by a British mathematician, G. H. Hardy (and independently in Germany by a physician, W. Weinberg). Hardy considered the problem trivial and in 1908 gave an immediate formula to show how genes are transmitted in populations, which he published in a short paper in *Science*. It never occurred to Hardy that this was of major significance to his colleagues in genetics or that his finding would lead to common ground for Darwinians and Mendelians. He made no mention of his genetics paper in his autobiography, *A Mathematician's Apology*. Weinberg's publication, some 20 pages long, appeared that same year in a German journal. It is today called the Hardy-Weinberg law and it is a founding basis of the field of population genetics.

The basic law requires nothing more sophisticated than junior high school algebra (which most readers of non-fiction who are not scientists have long ago forgotten). The two frequencies of a mutant gene (p) and its normal counterpart gene (q) among all the sperm in the population [p + q = 1] encounter, randomly, the same frequencies among the eggs of the population [p + q = 1], resulting in the simple equation $p^2 + 2pq + q^2 = 1$ where $p^2$ is the frequency of the homozygous dominant form of the gene AA; 2pq is the frequency of the heterozygotes Aa in the population; and $q^2$ is the frequency of homozygous recessives aa in the population. If a piece of the information is known, the entire equation can be solved. Thus, if 16% of the population carries a Red Cross card that identifies the holders as Rh negative, this means that $q^2$ is 0.16 (the number of Rh-negative blood donors in the population). The square root of 0.16 is 0.4 (40% of 40% is 16%), which is also the frequency of q. Since the population of sperm or the population of eggs consists of p + q =1, then p = 0.6 because 0.4 (the frequency of Rh-negative genes) plus 0.6 (the frequency of Rh-positive genes) = 1.0. Therefore, we now know the frequency of both genes among the reproductive cells that form the next generation. Since these genes will randomly encounter each other (e.g., you rarely know your partner's blood type when you begin dating and may not know it until you become parents), the resulting frequencies for AA is $p^2$ or 0.6 × 0.6 = 0.36. The heterozygotes, Aa, will be 2pq or 2 × 0.6 × 0.4 = 0.48. Knowing that the frequency of Rh-negative card holders is 0.16 carries some important implications. The first is that barring some type of advantage for one or the other of these traits, the frequency will not change from generation to generation. The population must be large and the geneticist assumes that new mutations are relatively rare and do not disturb this overall frequency. But evolution is different — mutations do occur, selection may favor either of the homozygotes or the heterozygote, and populations are not always large. Thus, the Hardy-Weinberg law, which is the source of stability that allows frequencies of genes in large populations in the short run to be estimated, becomes the basis (with all the factors that influence outcome frequencies) for evolutionary studies.

I enjoy reflecting on Hardy's dismissal of his contribution as so minor it would embarrass his reputation as a mathematician to have mentioned it in his autobiography. It tells me that not all major contributions to science are made with hard work, require complicated analysis, or lead to bragging rights by its discoverer. This is not because Hardy was shy or overly modest. It also tells me of the importance of synthesizing one field with another. Mathematics can be helpful to a relatively non-quantitative science like biology. Many people, including scientists in the physical sciences, are so dependent on mathematics to express anything in their field that they tend to dismiss less quantifiable sciences as irrelevant or not real science. Lots of experimental science is not mathematical. I also often reflect on this mathematical model of gene frequencies because it informs me of another important feature of science that is often misunderstood by non-scientists. A model is rarely identical to what occurs in nature. The Hardy-Weinberg law states that if a genetic disorder is rare, say albinos occurring one in 20,000 births, the frequency of recessives in future gametes of the next generation is the square root of that incidence which is 0.0071, and thus the frequencies of carriers (heterozygotes) is 2 x .9929 x .0071 = .0014, or about 1 in 714 people who carry the mutant gene for albinism but do not express it. Mathematics is a handy tool for determining such frequencies and informing relatives at risk or the general public about the chances of having a baby with a particular genetic defect. At the same time, it is slightly inaccurate because it ignores new mutations (from normal to albino), reverse mutants (from albino to normal), and the chances of failing to reproduce (the reduced fitness of a gene when it is homozygous or heterozygous).

Over a period of hundreds or thousands of generations (which is nothing to an evolutionist but often inconceivable to those of us whose visions of past or future are limited to three familial generations), this can cause a dramatic change in the frequency of genes in a population.

These four examples of mental inventions of scientists help us understand how rich science becomes as it makes associations from one field to another or provides mental tools for solving problems. These insights come from creativity, a process we recognize but have difficulty defining or teaching. At best, scientists can offer examples of those gifts to their students, and talented students will begin to mimic that inventiveness if they are lucky enough to be good at doing so. I firmly believe that bright students ignite each other's minds and this is why, over the years, I have tried to develop academic communities of bright students through courses, seminars, or activities that can bring about their creativity. I cannot teach creativity, but I can provide a climate that nurtures it. Alas, too often administrators lack vision and dismiss this possibility, although I have seen it happen in students since I was myself a high school student enjoying such opportunities with a few gifted teachers. Many of my colleagues, too, often feel that this approach is undemocratic and that bright students should not be concentrated in "elitist" programs. They fail to see that the unstimulated bright student loses out because the opportunities for intellectual self-stimulation or stimulation from unmotivated students and practical-minded teachers are remote. A gathering of minds, such as in Morgan's laboratory at Schermerhorn Hall at Columbia in the 1910s and 1920s, Delbrück's team at Cold Spring Harbor in the 1940s, or Watson and Crick at Cambridge where they could interact with scholars popping out of every door, is the world of Renaissance Florence or the world of Socrates' Athens writ small.

# Science as Scholarship

Most scientists read professional journals to stay abreast with their field. There are thousands of such journals and they are very expensive. Some libraries are charged several hundred dollars for a yearly subscription to a journal. The cost is high because most of these journals have few advertising revenues and few subscribers. There are only a few thousand universities in the world, of which only a few hundred can afford such subscription prices. As scientific journals have to be printed on paper that will last, the printing costs are substantial. Authors usually receive no payment for articles that are accepted, and referees who review submitted manuscripts serve without pay. Sometimes, the journal editors are themselves at universities and take on the responsibility without pay.

## Why Scientists Read Journals

To keep up, scientists usually limit their reading to a handful of journals. They may each read *Nature* and *Science* for articles that are noteworthy and that give a sense of how science as a whole is developing. Before my retirement, I read specialty journals in my areas of professional interest. As my field was human genetics, I regularly read the *American Journal of Medical Genetics* and the *American Journal of Human Genetics*, and these provided the best all-round coverage in that field. I also used to read *Fertility and Sterility* each month because I taught a course on the biology of human sexuality. I would walk through the medical library and pick up other journals, look at their table of contents, and check for titles that look interesting in *Lancet*, the *New England Journal of Medicine*, *Pediatrics*,

*Clinical Genetics*, *Nature Genetics*, *Nature Medicine*, and *Cell*. Being selective is probably typical of my colleagues even though their selection of specialty journals will differ from mine.

I may also differ from my scientific colleagues in reading *Harper's*, *The Atlantic Monthly*, *Skeptic*, *The Humanist*, *The New York Review of Books*, *The Nation*, *The New Yorker,* and *Archaeology.* I would read widely because my Biology 101–102 course was called *Biology: A Humanities Approach*, and I wanted to link what is going on in science to what is going on in our lives. But I note that many of my colleagues in the sciences do read journals that are not in their field, such as *The Economist*. When I served eight years as Master of the Honors College at Stony Brook, I had very little time to write books and articles. But I felt that it was important to be knowledgeable about many fields and I subscribed to fourteen different journals, which I displayed on a magazine rack that I had commissioned a talented premedical Honors College student to design and construct. I paid for this out of my own pocket because this was not something administrators care to subsidize, but it was something I felt that is important in creating an academic community.

## Scientists Vary in Their Interests Outside of Science

Although many popular books portray scientists as persons who have cauterized their souls and who act dead to the arts and humanities, I have not found this to be true. I hear classical music coming out of the doors of laboratories. I see many scientists at musical performances on campus. Most of my colleagues enjoy reading fiction and non-fiction. Most consider themselves scholars. They may not be professionally immersed in literature as are those in the humanities, but then I would not expect my colleagues in those fields to be deeply knowledgeable about the sciences. Doing well in any one field exacts a sacrifice in general knowledge. Some scholars, whether in the humanities, arts, sciences, or social studies, do become very narrow and lose their capacity to relate to colleagues in other fields.

Scholarship is something akin to being well-read. Scholars love to read because this informs them about what is new in their areas of interest. Things were different when I was an undergraduate. Undergraduates were not allowed into the graduate library and they did not read professional journals. I remember my shock when I was a first-year graduate student at Indiana University. The journal articles were virtually so opaque that I could not understand them. There was so much expectation that I should know techniques and results of work in the field, and yet too many terms I did not know. Nothing was reviewed and nothing made sense — there was hardly any entry even at a beginner's level. This experience was also true of the graduate seminars given by my fellow graduate students. The first year I sat in silent terror, afraid that I would be unmasked as a fraud. I grasped not much more than ten percent of what I heard. However, by my third year, I understood everything and I could read professional journals in my areas of interest with ease. By the time I was writing my dissertation, I knew the professional literature in my field very well. I would pick up an article and read not only what was in it but also what was left out because each investigator has preferences for certain ideas, approaches, or papers and ignores material that is considered irrelevant, untrustworthy, or trivial. At this level, the scholar has acquired critical judgment. Hence, the scientist reads for both the information and the significance of the author's work. If there are glaring errors, unfounded omissions, or careless ways of presenting the material, the reader spots it.

# The Importance of Narrative in Discussing Science

Scholarship manifests in a variety of ways for the scientist. Instead of reading to gain knowledge and a steeping in one's field, the scientist may read widely to write about the field. Reading a lot provides massive amounts of information and some scientists take time out from conducting experiments to review what is going on in the field. Broader theories may sometimes emerge from the connections achieved by reading widely, leading to reviews in the form of articles or books.

Muller wrote several such scholarly articles, three of which had wide influence. He wrote one on *Variation due to change in the individual gene*; one on *The gene as the basis of life*; and one on *Life*. None presented any new data or summarized experimental data. They were broad scholarly papers that awakened his audience to new ways of looking at material that had been out for several years. In *Variation due to change in the individual gene*, Muller pointed out that biologists in the past considered heredity and variation to be distinct phenomena. Before the 1920s, heredity was considered to be the inheritance of species characteristics or overall likeness, while variation was considered to be fluctuating departures from heredity. Heredity was considered fixed whereas variation, as the name implied, was considered subject to modification by the environment. These ideas would have seemed perfectly sound to Darwin and his contemporaries. There was no doubt about how Muller learned his biology and even how the first text in modern genetics, R. H. Lock's *Heredity, Variation, and Evolution* (1906), accepted that viewpoint. Muller presented his paper on variation in 1921 at a meeting in Toronto. He began: "The present paper will be concerned with problems, and the possible means of attacking them, than with the details of cases and data." He developed his case for the importance of the gene as the unit of heredity and how it affects every process, structure, and system in the organism. He pointed out the unique feature of the gene, which is its ability to copy its own errors and still retain the fundamental property of being a gene. He rejected using the term mutation for anything but changes in the gene, providing a new vocabulary for structural changes and for gains and losses of chromosomes. This change shifted the focus of scientists toward the individual gene and its significance. Although Muller admitted that neither he nor any other scientist could propose a successful model of the structure of the gene and how it works, he had already immersed himself in almost a decade of work in fruit fly genetics and had some insights to share. He pointed out: "Inheritance by itself leads to no change, and variation leads to no permanent change, unless the variations are themselves heritable. Thus it is not inheritance *and* variation which bring about evolution, but the inheritance *of* variation, and this in turn is due to the general principle of gene construction which causes the persistence of autocatalysis despite the alteration in structure of the gene itself." After reviewing what geneticists knew about mutations, mutation rates, and the changes taking place in individual genes, he proposed that geneticists should study newly discovered bacterial viruses because their properties are strikingly like genes. In those days, bacteriophages were called "d'Hérelle bodies" after their discoverer, Félix d'Hérelle (1873–1949). Muller revealed a bold vision: "…if these d'Hérelle bodies were really genes, fundamentally like our chromosome genes, they would give us an utterly new angle from which to attack the gene problem. They are filterable, to some extent isolable, can be handled in test tubes, and their properties, as shown by their effects on the bacteria, can then be studied after treatment. It would be very rash to call these bodies genes, and yet at present we must confess that there is no distinction known between the genes and them. Hence we cannot categorically deny that perhaps we may be able to grind genes in a mortar and cook them in a beaker after all. Must we geneticists become bacteriologists, physiological chemists, and physicists, simultaneously with being zoologists and botanists? Let us hope so."

## Some Scientists Prefer to Write Books

Muller preferred writing articles as he felt that books were too demanding of his time. He could write an article in a day or two, but it would take him months or years to write a well-documented or well-reasoned book. He wrote a popularization of genetics and society in hopes of arousing interest in his own eugenic views for the future in a society that was free of bias, which was a rather naïve conception but not surprising for an idealist like Muller. Other scientist scholars have pursued a successful career writing books. One was a monograph written by E. B. Wilson (1856–1939), one of the founders of the chromosome theory of heredity. He was also the chair of the Zoology Department at Columbia and hired Morgan to develop experimental biology there. Wilson is best known for clarifying the sex chromosomes, which were confusing, controversial, and puzzling when the story was emerging. Wilson, like Morgan, got his PhD from Johns Hopkins University, and preceded Morgan in making the grand tour of Europe and working in the best laboratories of Germany before returning to the United States. He enjoyed microscopy and spent most of his professional life studying chromosomes and cell components through careful observation. He was also a synthesizer who could see order among hundreds of findings by his colleagues. His lectures never rambled; they were polished and always had a theme, and he excited his students with the latest results from his readings. He put together his lecture notes and polished them into chapters to produce a book in 1896 titled *The Cell in Development and Heredity*. It became the vanguard for the new experimental science and high quality investigations carried out in Europe, inspired by the German university system. Wilson recognized that the study of chromosomes in cells was leading to a new theory of heredity. The nucleus of the cell and its contents would be the focus for such studies because cell division had been worked out, the chromosomes seemed to have different effects when one or more was accidentally lost from a cell and led to the formation of an embryo, and the precision of cell division made no sense if it were not somehow associated with preserving the hereditary specificities of the organism.

Wilson marshalled evidence for the growing belief that nucleic acids were the chemical components of the chromosomes. Sperm nuclei were virtually pure nucleic acid and the sperm contributed half the heredity at fertilization. When a nucleus was removed from a cell, it ceased making substances needed for its survival and exhausted what it stored within a few hours. Thus, Wilson could conclude that nucleic acids "may be in a chemical sense not only the formative center of the nucleus but also a primary factor in the constructive processes of the cytoplasm [...] for the recurrence of similar morphological characters must in the last analysis be due to the recurrence of corresponding forms of metabolic action of which they are the outward expression. [...] This presumption becomes a practical certainty when we turn to the facts of maturation, fertilization, and cell division. All of these converge to the conclusion that the chromatin is the most essential element in development."

The vocabulary and phrasing of 1896 may differ, but Wilson's essential argument is clear — by 1896, there was sufficient evidence to identify the nucleus and its chromosomes as the source of those metabolic activities that made the cell live and develop into an organism. The chemistry had been sufficiently advanced for him to feel confident that nucleic acid was the major substance composing the chromosomes and thus the factors associated with heredity and its expression in development and throughout the life cycle. If one considers the student audience in 1896, this must have been electrifying because zoology and botany were largely descriptive and far removed from chemistry. Now it was possible to think of heredity as having a cellular basis in the chromosomes as well as a chemical basis in the nucleic acids.

A very different book was offered in 1945 by Erwin Schrödinger. He called his work *What is Life?* Schrödinger had left Austria after the union with Germany and its occupation by the Nazis made it difficult for any intellectual who was neither Aryan nor sympathetic to Nazi ideology to continue work. He found refuge in Dublin and continued his studies of physics there. His only obligation was to prepare and present a series of Christmas lectures for the public, a tradition that many universities in Great Britain had continued since the late nineteenth century. Schrödinger's claim to fame was his development of wave mechanics which established him among the major physicists of the first half of the twentieth century. Like Delbrück, he was strongly influenced by the emerging molecular biology which was bringing physics and biology into a common inquiry. He had read a provocative paper published in 1935 by Delbrück and colleagues that attempted to develop a quantum model of the gene and used much of Muller's findings on radiation-induced mutations. The paper also attempted, not too successfully, to measure the gene using target theory, a shooting-in-the-dark approach that sometimes worked for estimating sizes of molecules or atoms in a known volume.

Schrödinger began by raising a question about chemistry, physics, and biology. The so-called exact or physical sciences use statistics or at least what can be considered a statistical outcome for their laws. When a spoon of sugar is placed in a cup of tea, the number of molecules of sugar or the number of molecules of water is truly staggering to the human imagination. It is of the order of 10 to the 23$^{rd}$ power, approximately what is called Avogadro's number, which is beyond anything familiar to us. Probably the largest number we encounter in social life is about one to ten trillion, roughly the amount of money in dollars there is in the United States. A trillion is a one followed by 12 zeros. If the US accounted for only one percent of the world's wealth, which is a very conservative estimate, the world would be worth one quadrillion dollars (ten to the 15$^{th}$ power). This number of molecules would be an almost invisible or undetectable amount of sugar in a spoon. When that sugar is stirred in the tea, the molecules quickly become dispersed. Here, the statistics begin to make sense to us. It would be virtually impossible for all of the sugar molecules to move to one spot in the cup. Each sugar molecule would collide with other sugar molecules as well as with water and ions in the water, and these collisions would move some molecules up, some down, some to the left, some to the right, some forward, and some backward, as well as many in all intermediate directions. This is called Brownian motion. If we think about it, the motions will be randomized and no one part of the fluid volume would have more or less sugar molecules in it than any other part. If the path of one such sugar molecule were followed, it would appear to jitter about in a drunkard's path. Whatever neighboring molecules it had would soon separate from it. Yet biological systems are dependent not on a huge swarm of genes working like sugar molecules in a solution, but rather existing as individual genes. In a sperm or an egg, virtually every one of those genes is unique. What keeps them from flying off from the Brownian movement collisions that characterize the motion of dissolved sugar in water? How does the activity of a single gene amplify its way up from the individual molecule to an organism? Somehow the gene pulls it off.

Schrödinger made several inferences that were partly what Muller, Delbrück, and his colleagues had surmised about genes from their biological properties. Schrödinger tried to convert those ideas into ways that a physicist would look at biology and proposed that the gene was a special kind of crystal, which made sense because the gene copies itself and that's what crystals do very well. But most crystals are boring; they copy a repeating pattern. The repeating pattern is called periodicity and thus most crystals are periodic. For table salt, it would just be a stack of alternating sodium and chlorine atoms. Schrödinger was aware of the problem Muller raised: genes not only copy themselves, they copy their mutant variations. Thus, genes are aperiodic crystals; they lack periodicity.

Schrödinger pursued the gene in chemical terms. No one knew, when he wrote his book, whether genes were proteins, nucleic acids, or something else. But to Schrödinger, this was not essential. Whatever chemical composition the gene had, it had to have an additional property that specifies or brings about specific hereditary traits. Schrödinger suggested that genes used a molecular "code script." This code need not be elaborate because, as Schrödinger pointed out, we can code the 26 letters of the English language with a Morse code consisting of three symbols — a dot, a dash, and a space. Schrödinger made one error in his analysis when he considered the organization of living things. Well-organized systems tend to break up and decay into randomized disorder. As they do so, their entropy increases. Hence, in the physical world, ordered structures sooner or later become disordered and entropy increases. Schrödinger could not understand how life maintains its order. He suggested that life is governed by some new law of physics that generates "negative entropy," which is a mystical concept and almost invokes some connection to the divine, a thought that did not escape Schrödinger. Most biologists would explain the maintenance of order as one that comes at the expense of a trail of positive entropy as we metabolize our foods, inhale our oxygen, and discharge wastes such as gasses, urine, heat, and other signs of entropy. The total entropy is always positive in living systems and it is the metabolism of the cells that keeps the order going through the life cycle. Life requires a constant pouring in of energy to maintain itself.

The inspired use of the term "aperiodic crystal" found its realization in the aperiodic sequence of nucleotides that characterizes each unique gene. Similarly, the inspired term "code script" found its expression in 1960 when the genetic code was deciphered by chemical means. We now know that that code is universal for life on earth, with some 20 amino acids encoded by 61 of our 64 possible triplet combinations (called codons) formed by nucleotide sequences. The remaining three codons specify stop signals, just as a space between words tells us when one word ends and another begins.

Many physicists, chemists, and mathematicians read Schrodinger's *What is Life?* Some joined the ranks of molecular biology and took on the analysis of bacteriophage genetics while others, like Crick and Watson, applied their skills directly to uncovering the holy grail of the DNA composition of genes. It is rare for a scientific book to change the direction of a field. Darwin certainly did so with his *Origin of the Species*, and it is even more remarkable that a physicist with no formal training in the life sciences, but with the power of a scholar's critical mind, did the same for genetics.

# Science as Experimentation

Experimentation is almost a synonym for science in the public mind. School children were taught about experiments when they went to class. They learned a formula called the scientific method and that every experiment has a control. Each experiment teaches us something about the one variable that produces the experimental outcome. These ideas are not wrong, but science does not comprise only of experiments. We have already seen the importance of discovery, exploration, and invention, and there is much more as we shall see. But experimentation captures our imagination like no other aspect of science does because of how it was emphasized during our early school years as well as the many examples that exist, from smashing atoms to showing that the hereditary material was indeed made of DNA. All of us remember the tedium of preparing laboratory reports for our high school or undergraduate science courses. The methodology drummed into us made it almost a religious ritual. Moreover, it often had not so desirable consequences in some students — not an increased appreciation of science but a contempt for it and a temptation to hand in to the instructor what the student thought the instructor wanted: not the actual data but the data that fitted the predictions and all too obvious outcome.

## Demonstrating the Germ Theory is a Classic Example of Experimentation

The easiest experiments to grasp are those where the effects of products are tested. In our imagination, we can easily see a Louis Pasteur or Robert Koch injecting 50 sheep in one field with anthrax grown in the laboratory and then injecting another 50 sheep in a nearby field with an equivalent amount of sterile broth that is used to grow bacteria. We can come back two or three days later to find 50 dead sheep in the first field and 50 live sheep in the second field. The skeptics are convinced and they applaud our hero. More specifically, the control (not infected with the anthrax culture) convinces us. It is not injection, but what is injected, that caused the disaster to one group of sheep. The sheep, the fields, and all other conceivable conditions are the same except for what was injected. Thus, our imagination allows no other interpretation. Without the control, doubts would linger. This experiment, in our minds, is a success because it was done right.

## Experimental Results Often Require Statistical Analysis

All or none responses of this sort are not very common in science. Often, success is measured statistically. For instance, the treated batch of animals (often a smaller number, such as 15 or 20 mice or rats) may show that two thirds of them slow down when fed a new sedative. The control batch may show about half slowing down when fed a placebo (the fluid in which the drug is dissolved, but not the drug itself). How does one assess 10 out of 15 compared to 7 out of 15? Suddenly, the scientist as experimenter shifts to the scientist as statistician, and another formula enters the story. This formula is some sort of mathematical way of adding, subtracting, and using differences, squaring differences, and contrasting these differences to predicted outcomes and seeing if they match. It also involves determining whether repeating the experiment 95 out of 100 times will get you similar results, which means that the results are significant. If you fall below that desired 95%, then the results are inconclusive and it's time to do the experiment again (possibly on a larger scale) or try another chemical modification of the drug.

To make things seem even fairer, another wrinkle is thrown into the experiment — you won't know which animal got the drug and which got the placebo. Someone else will prepare the control injections and the drug injections. You will only know a code number that is known to the outsider who prepares the injections. Only after the results are in (sometimes scored by yet another outsider) can the sealed information be opened for you to find out which organism got the drug and which got the placebo. These "double blind" experiments are often used to prevent self-deception and errors that reflect bias — conscious or unconscious — by the investigator.

For many a scientist, these precautions are not needed because the design of the experiment yields qualitative results. You know what is happening because you actually see a difference, just as the dead sheep drive a message home. Scientists also know that if the results are good and the experiments are published, others will repeat the work. An experiment has to be written up in a careful and detailed manner so that someone reading the published paper can repeat, extend, or modify the experiment to either confirm it or explore other implications. Thus when Muller, using fruit flies, and Lewis Stadler, using maize, both reported the induction of mutations with x-rays in 1927, many investigators around the world repeated the results in dozens of different organisms. Very few would have reconstructed Muller's or Stadler's experiments and repeated them item for item. Doubts would have arisen if x-raying at these doses did not induce mutations in snapdragons, wasps, grasses, mice, or beetles. Most of those investigators could not have duplicated Muller's approach, which used carefully designed genetic stocks to identify a specific class of mutations (X-linked lethals) that damaged genes on the X chromosome. Most confirmations came from the search for visible mutations in the second or third generation after the original exposure to radiation.

## Not all Findings in an Experiment can be Interpreted at the Same Time

Muller's paper, which was presented at the 1927 International Congress of Genetics in Berlin, was exciting because of its cleverness in experimental design. Each mutant that Muller obtained, including all the lethals, were mapped. He also explored whether the later generations could "throw off" mutations, reflecting some sort of instability in the genes or a compound nature of the gene like a bean bag. He did not find this and instead found a possible "doubleness" of the chromosome in the sperm because of a high number of induced mosaic mutations that sorted out into mutant or non-mutant offspring. This doubleness remained a puzzle for another thirty years until the double helix composition of DNA led to an explanation for mosaicism. Muller also tested for differences between x-rayed sperm and x-rayed eggs and found a higher mutation rate in the sperm. The great excess of recessive X-linked lethal mutations to X-linked visible mutations proved that mutations that act during the embryonic stages of the fly are far more numerous than those that lead to viable organ defects in the adult. He showed that the induced visible mutations were similar to those that had arisen spontaneously over the previous 15 or so years, which proved that x-rays produced the same kinds of changes that normally occur much more rarely as accidents. By 1931, he was able to demonstrate that the total amount of radiation in the surrounding environment or from cosmic rays only produced a minute fraction of one percent of spontaneously arising mutations and, thus, most of those mutations were probably caused by chemicals from the body's own metabolism. What distinguishes ordinary experiments from brilliantly designed and executed ones is the richness of ideas and findings that emerge from them.

## Chemicals Differ from X-Rays in How They Produce Mutations

Charlotte Auerbach greatly admired Muller. She was a refugee from Nazi Germany and found an opportunity to continue her graduate work in Edinburgh. Muller had just arrived in Edinburgh after the collapse of the Spanish Loyalists in their war with Generalissimo Franco who had the fascist support of Germany and Italy. Muller was told by the director of the Edinburgh genetics program that he would be given a postdoctoral student to do his cytogenetic work. When Muller interviewed Auerbach, he quickly learned that she was grateful to work for him but cytogenetics was not her interest. Muller asked her what she really was interested in and she mentioned studying the gene. He gave her a free hand to find a topic of interest and she began to study chemical mutagens. At that time, there were none — Muller himself had tried several compounds (e.g., vital dyes) without success in the 1920s. Auerbach tried her hand with tars similar to those produced by cigarettes because some of them were known to cause cancers when smeared on shaven rats. There was a suspicion that cancers may have some origin in defective genes. The outbreak of WWII led Muller to return to the United States and Auerbach was now on her own. She followed a lead given to her by a pharmacologist who suggested using mustard gas or nitrogen mustard because these compounds produced damage similar to overdoses of x-rays. She began spraying containers with these agents and her own hands began to blister and became raw from the chemical damage to her skin. But she got "heaps" of mutations using similar stocks that Muller had used to detect mutations. War time secrecy prevented her from publishing her results until 1946. She did write to Muller about her discovery, but she could not mention the agent that she had used. Auerbach, like Muller, founded a field. She noted many differences between chemical- and radiation-induced mutations. Chemicals produced more visible mutations, many of them mosaic when they first appeared. She recognized that treated sperm was far more likely to produce mutations than the same doses applied to eggs (the huge wads of cytoplasm in the eggs absorb and neutralize the chemical mutagen before it reaches the chromosomes in the nucleus).

Another implication of successful experiments is that they lead to extensions in somewhat different directions. The experimental design that reveals x-ray damage can also be used to reveal whether chemical agents act as mutagens. Dozens of chemical mutagens have been found this way. Sometimes the agents are applied as a gas or spray as in mustard gas, sometimes the flies are fed the agent (the alkylating agent ethyl methane sulfonate is a good example), and sometimes the agent is injected. The most unusual method that I can recall as a graduate student was a technique used by Irwin Herskowitz, Muller's laboratory associate. He used a vaginal douche technique, squirting the agent into the vagina of fruit flies after they had copulated with fertile males. My own experience using chemical mutagens was instructive. I was at UCLA and decided to induce mutations with chemicals using a specific gene, the one that I had studied for my dissertation where I induced mutations with x-rays and mapped them along with those that arose spontaneously over the years. I flew to Philadelphia and was trained by Irwin Oster, one of Muller's students who had learned the techniques from Auerbach. I used a quinacrine mustard and my first efforts at injecting flies were awful — most of the flies were butchered by my awkward technique! As a graduate student, I had plenty of doubts about my own manual dexterity. I had taken a laboratory course in experimental embryology with James Ebert (later President of the Carnegie Institution of Washington), and most of the eggs whose shells I dutifully sawed to create a window through which I could observe the development of chick embryos soon fouled into rotten eggs. Ebert told his graduate students, "Elof would have difficulty dissecting fetal elephants."

But I learned that practice makes perfect, and when I returned after injecting vital dyes for practice, the flies lived. I then injected enough mutagen to distend the abdomens of the male flies, mated them up some 24 hours later, and looked for mutations in the dumpy gene I was so interested in. Like Auerbach, I got heaps but about 97% were mosaic. This finding was quite a surprise and it led to a nice developmental study on how much of the mutant tissue in those flies entered the reproductive tissue and got transmitted. It also suggested to me that the chemical differed from radiation in its effects on DNA. I surmised that x-rays spread so much energy that the nucleotides were splattered from direct hits, but the chemical agent merely attached itself to one side group of a nucleotide base which led to that strand generating a mutant and the unaffected strand of the DNA producing a normal gene copy.

## Each Experiment has the Potential to Stimulate New Experiments

The publication of a good experiment creates a continuity effect — the work keeps on going. Muller designed the stocks so that these experiments could be done, and he also showed that they worked for x-rays. His postdoctoral student used it for chemical agents and showed that these too produced mutations. She then taught one of Muller's graduate students many years later how to do chemical mutagenesis, and I flew out to Philadelphia some years after his trip to Edinburgh and learned how to use injection techniques. In each case, the experiments differed in stock design, the types of problems explored, and the implications of findings. The story becomes even more impressive when Milislav Demerec read of Muller's work (which, as a lesser feature, disproved his model of the gene as a bean bag). Demerec began using the radiation technique to study new problems while at Cold Spring Harbor in New York. One of these problems arose in the 1940s when culturing penicillin was attempted by Florey and Chain in Great Britain, and they found that the antibiotic could be produced albeit slowly in a mold. Demerec was interested in the new molecular biology using microbial systems and he tried irradiating the mold with x-rays. To his delight, and within a few months of effort, he obtained mutant strains of *Penicillium* that increased the output of the antibiotic by one hundredfold. The commercial scale production using these strains changed medical history. Instead of young men and women dying of pneumonia, a shot of penicillin rescued their lives. Medicine had shifted from handholding and easing patients into their death to actual cures using agents that worked.

## Money and Talent do not Always Produce Positive Results

Most of the public perception of science is topsy-turvy. The public (and often government legislatures) believes that if science sees a problem out there, it can use some skills that include the scientific method to solve the problem. This view is based on the social belief that money applied to problems solves them. Often it does, but to scientists, most of the problems that society wants solved do not come from a direct confrontation to solve it, but instead from the applications of pure science many years later. Thus, Muller never considered his experiments with radiation as a means of making more milk per cow or more eggs per hen, or making genes yield commercially useful products. He was interested in basic questions: What is a gene? How big is it? What is its shape? How often does it mutate? What kinds of changes do you get if you keep getting mutations from a particular normal gene? Can one induce mutations? How does a gene work? Scientists ask governments to help them do basic science because they know that in unexpected ways the findings of basic science lead to practical results.

## Experimental Embryology has its Origins in Nineteenth Century Germany

I will shift from genetics to developmental biology because this too will give us insights into the way experiments are done. Experimental embryology was largely an invention of German and other European scientists in the late nineteenth century. Why did it develop there and not, say, in the New World where there is less tradition and more freedom to do what we want? Germany invented the modern university. In the late eighteenth century, scholars like Goethe and the Humboldt brothers promoted the idea that science, all of knowledge, and the scholarship that produces it is a public good. The German university was built to train scholars, not lawyers, doctors, engineers, musicians, or businessmen. The applications of knowledge was secondary to the creation of new knowledge. The business of professors was to explore the unknown, publish their findings, and teach young men (women were excluded in those days) how to become scholars. Students were given academic freedom, which meant that they could choose to attend classes or not, go to any university in the German states to take courses, and come back when they were ready to defend their reputations as scholars. There were no examinations in courses, but there was a panel of professors who met at a student's request. Students were grilled on their knowledge in their field and on the original work they did as scholars. The Germans created a professorial degree, the PhD.

Americans did not pursue that route until 1875. Before then the American college was primarily a place to produce Christian gentlemen. It provided a liberal arts education and some testing of the tenets of one's faith to certify that these were good citizens who could then apprentice themselves to MDs or lawyers and become practiced in their art, or who could establish reputations for themselves as business leaders in their communities. In 1875, Daniel Coit Gilman (1831–1908) came back after a tour of Europe with a vision to use the money given him to found the Johns Hopkins University. Hopkins wanted something new in American education, so Gilman combined the ideals of the German university with the practical knowledge of British working class colleges. He spoke to utilitarians like Thomas Henry Huxley and created a graduate school where faculty would do the top scholarly work found in European universities while developing an under-graduate program to steer potential scholars into graduate programs.

In the European university, MDs and PhDs learned that experimental medicine and experimental zoology were beneficial in revealing how the body works down to the cellular level and that from basic science, practical applications might arise. Some scientists found that description was not enough and that it was possible to manipulate the cells and tissues of marine organisms and learn fundamental things about life. Hans Driesch (1867–1941) was one such scientist. He used sea urchins and found that he could separate an early embryo at the two-cell stage into separate cells. Each cell went on to form an identical twin. This was a shock to him as he thought that he would end up with abnormal embryos. Driesch's interpretation was that each cell was "totipotent," or that it retained the potential for full embryonic development. At some later stage, presumably, cells lose their totipotency and become differentiated. Such cells would abort if removed and cultured. Wilhelm Roux found a different story. If he cauterized one of the two cells of a frog at this stage of development, a half embryo formed. He believed that the two cells were not totipotent and that this represented some sort of species difference. When young T. H. Morgan came to Europe and worked in Driesch's laboratory to learn experimental embryology, he found that he could separate the frog cells without burning one off. When the two cells developed, each formed a healthy frog. The two cells were totipotent. This meant that the damaged material of the cauterized cell interfered with the development of the intact cell. Morgan also found that he could take two fertilized eggs from a salamander or some other suitable organism and fuse them together to form a single salamander. It was the reverse of twinning — what should have been two individuals had become fused into one individual, a chimera. Much of Morgan's work was done in Italy, near Naples where a consortium of European scholars purchased a building and made it into a marine biology station for research. Universities or individuals paid for the use of a table to do their research and they enjoyed the intellectual excitement of getting together to talk about their work. The leading figures in founding this station were a German Professor (Anton Dohrn) and an English Professor

(T. H. Huxley). Morgan was well known to the Italian community because his father had helped Garibaldi unify Italy.

## T. H. Morgan Shifts from Embryology to Genetics and Launches a New Field

When Morgan returned to the United States, he carried out experimental zoology first at Bryn Mawr and then at Columbia University where his fellow Hopkins alumnus, E. B. Wilson, was now Chairman of the Zoology Department. Morgan chose the title, Professor of Experimental Zoology. His switch to genetics was partly inspired by a later trip to Europe when he visited Hugo De Vries's laboratory in Holland. De Vries was one of the rediscoverers of Mendel's work and De Vries believed that he had found new species arising out of the evening primrose (*Oenothera lamarckiana*). This led Morgan to hunt for new mutations in a suitable organism, including mice, guinea pigs, and pigeons, but on recommendation by a colleague at Harvard, he began to breed fruit flies for their convenience. He did not find new species, but he did find mutations.

## Mouse Embryos Provided Insights into the Process of Organ Formation

Morgan dropped his major emphasis on developmental biology and con-tributed greatly to classical genetics. Later in life, he returned to embryology but did not make any major contributions to the field. A major contribution to experimental embryology arose when techniques for studying mammal embryos came about, which is important because we are mammals and what is found in mice may be of more significance to us than what is found in a marine annelid, fish, or frog. The major experiments worthy of our attention were developed in the 1970s in England and in Philadelphia. Anne McLaren developed techniques at Cambridge to cut up early embryos and switch parts. Beatrice Mintz developed techniques to commingle the cells of early embryos, producing something like what Morgan had produced — single individuals from the fusion of potential twins. Each of these sets of experiments revealed a lot about the stages of the embryo that are somewhat later than those that Morgan and his colleagues could manipulate at the turn of the century.

McLaren concentrated on a stage of the mammal life cycle called the blastocyst, which is a structure that forms from a totipotent ball of cells and enters the uterus. The ball forms a complex structure with an outer layer of cells, the trophoblast, and an internal smaller ball of cells, the inner cell mass. The trophoblast is the tissue that implants into the uterine lining. The inner cell mass later differentiates into embryonic layers from which all the organs are formed. Every part of your body was derived from the inner cell mass; none of it was formed from the trophoblast. How do we know that this is true? McLaren first tried cutting a blastocyst into two. In one cut, a vertical slice gives two half blastocysts, each half with some trophoblast and some inner cell mass. The two pieces each can form a healthy twin mouse. If she cut it horizontally so that one piece was all trophoblast and the other had a reduced trophoblast but all the inner cell mass, the latter forms a mouse and the empty trophoblast implants but aborts. It produces a false pregnancy and there is no embryo that forms.

McLaren then tried switching inner cell masses of two different strains (one albino and one pigmented). Invariably, the mouse that formed showed only the genetic characteristics of the inner cell mass donor and none of the host trophoblast. This outcome was also true if she injected cells into the inner cell mass. If it were a pigmented cell from an inner cell mass inserted into an albino inner cell mass, the resulting mouse had a splotch of pigmented fur. If she injected a pigmented trophoblast cell, however, the recipient albino inner cell mass produced a perfect albino mouse. The trophoblast could not become part of the embryo or adult. It had lost its totipotency.

Mintz tried something different. She squashed together two mouse embryos (one albino, one pigmented) in the eight cell stage and the resulting sixteen cell stage produced a blastocyst with a mixture of albino and pigmented cells. Such mixtures produced a chimeric mouse, often patchy like a crazy quilt. Usually they were male or female and not mixtures of both sexes, but a few did show up as hermaphrodites. She inferred from this that sex determination in mice is somewhat different from that in humans, where chimeras are usually intersexed with internal genitalia that are both male and female and the external genitalia are ambiguous. She also used a variety of mutations in the mice and commingled these cells with normal cells of the same stage to see if she could determine how many cells go into the initial formation of an organ. Some required only one or two cells (like the ducts that form the uterus or the ducts that form the sperm transport pathway from the testes to the penis), others required about a half dozen to a dozen cells, like the retinas of the eyes. This information is intriguing because it offers insights into how genes work and when they work. It is technically much more difficult to dissect mouse blastocysts and inject cells or to commingle eight cell stages and reimplant these in a uterus than it is to dissect relatively large frog eggs. A frog egg is thousands of times larger than a mouse egg.

## The Findings of Experimental Embryology Lead to Challenges in Philosophy and Theology

The work of McLaren and Mintz raised fundamental philosophical questions about what constitutes an organism. We do not doubt, for example, that identical twins are not one but two separate people. But what is a person who was supposed to be non-identical twins? Is this chimera also a single person despite being the product of two separate eggs fertilized by two separate sperm? What sex do we assign to a hermaphrodite? If it is a mouse, that would not be much of a legal or philosophic issue, but for a human it can be a very difficult one because what is put on a birth certificate will follow that person for life and, in all likelihood, determine how that person will be named and raised. Experimental embryology provides a model of how a human birth defect takes place. Experiments performed by William Castle at Harvard on rabbits shifted early albino embryos from one mother's uterus to a foster uterus of a black furred rabbit. The experiment showed that there was no influence on either the resulting guest rabbit curled in its mother's womb or on its offspring when it was later mated to an albino rabbit. It was the genetic constitution of the embryo that mattered, not the maternal nurturing environment, for basic biological traits such as fur color. The experiment also showed that it was possible to do experimental embryology on mammals, but that was not taken up again for some 50 years. The techniques to do experiments on mammalian cells had to catch up with the ideas. In our own time, experimental embryology has given us the field of in vitro fertilization and remarkable techniques to help sterile people become parents. Neither Driesch, Roux, Morgan, nor Castle would have predicted such an outcome from their work.

# 6 Science as Theory

Some scientists are brilliant theoreticians and some are superb experimenters. A few are both. Both approaches are essential for science. I once invited Nobelist Chen-Ning (Frank) Yang (b. 1926) to speak about his life and career to students in one of my courses. He described how his father was a mathematician and how he grew up in Amoy until the Japanese invasion forced his family to move to Kunming where the Burma Road ends in southwest China. He went to a school that was little more than a hut with a tin roof and earth as the floor. When it rained, the pinging of the rain on the roof reverberated all day through classes. He showed a picture of himself and his classmates in rags. He was a good student and highly motivated to learn, as were his classmates. Yang had thought of a career as a chemist, and one of his teachers who had once attended the University of Chicago told him that when the war was over, he should apply there. When that occasion came, Yang was already interested in physics. He had the good fortune of studying with both Edward Teller (his dissertation advisor) and J. Robert Oppenheimer. Oppenheimer then left for Princeton to head the physics program at the Institute for Advanced Study. Although Yang liked what he did for his dissertation, he admired the work that Oppenheimer was doing at Princeton even more, so he went to the Institute with the "young Turks" to work with Oppenheimer. Oppenheimer reasoned that young minds were more likely to have fresh ideas, and he also believed that having a cluster of young physicists to talk about their work all day long was the best way for them to come up with innovative ideas. He was right. Yang, with his colleague, Tsung-Dao Lee (b. 1926) came up with a theory that suggested that the time-honored principle of symmetry in the universe (everything has its opposite counterpart) was wrong. The so-called conservation of parity turned out to be flawed on theoretical grounds. Within a few years their theory was tested and the predicted exception to parity was found. The universe is intrinsically asymmetrical. Yang and Lee shared a Nobel Prize a few years later.

Yang told us that he tried his hand at experimental physics while he was a graduate student and learned that he was clumsy and dangerous. The motto of his fellow students was, "If there's a bang, it must be Yang." Yang headed the Institute for Theoretical Physics at Stony Brook until he retired and returned to China. Physics as a field lends itself to full-time theorists and full-time experimentalists. This composition is less true in the life sciences where there are few full-time theoretical biologists. One reason for this difference resides in the relative universality of things in physics (about 100 elements, a fixed number of particles, about a half dozen major laws, and the statistical nature of the universe at the macroscopic level where huge numbers of atoms or molecules obliterate variation). In the living world, differences and uniqueness predominate among species, organisms, and their parts. It is difficult to have universal biological laws when there are so many variations, which arise from the constant mutation of genes and rearrangements of chromosomes over thousands of generations of life separating species.

## Theoretical Biology is Less Common than Theoretical Physics

Among the great theoretical biologists was August Weismann (1834–1914), who spent most of his professional life in Freiburg in Germany. As was customary in those days, the MD was a scholarly degree requiring research and a dissertation. Weismann took an interest in jellyfish (hydrozoa) and flies (diptera). He would no doubt have continued his work as a microscopist had not failing eyesight and functional blindness assailed his career. During periods when he could neither see through a microscope nor even read journal articles, he had his students do the observations for him and he dictated his own articles. These sieges of blindness came and went, often relieved for a short period of time at ten year intervals. From 1864 on, he was essentially a theoretical biologist.

Weismann's first great contribution arose out of animal studies. He realized that germinal tissue was set aside. In flies, the egg that is laid is actually an embryo and the future germinal material proliferates in a segment of the embryo called a polar cap. Later when the future ovaries and testes form, the polar cap cells migrate and invade them. Weismann argued that this must reflect a fundamental difference between multicellular organisms and unicellular ones. The unicellular organisms appear to be immortal because they divide and you now have two cells from one cell with no death except if victimized by environmental circumstance. There did not seem to be a life span for such protozoa and algae. With multicellular plants and animals, however, the setting aside of reproductive cells (which he called the germplasm) permitted the rest of the organism (which he called the soma or somatoplasm) to live without reproducing for a fixed duration called length of life or life expectancy. It is only the germinal cells that are transmitted to future generations and not the somatic cells. This theory of the germplasm had an important implication. He argued that changes in the soma did not require corresponding influence on the germplasm. Whatever variation that existed was due to spontaneous or transmitted variations in the germplasm. Variations occurring in the soma had no way to reach the germplasm. If this were true, Weismann argued, then the belief — widely held since antiquity — that the environment directly alters heredity in a predictable direction is wrong. It was Jean Baptiste de Lamarck (who coined the term "biology" in 1801) who had advocated the theory of transmission of acquired characteristics to interpret the evolution of life, and to this day his erroneous theory is called Lamarckism.

Weismann published many monographs on the theory of the germplasm and its implications. He attempted to disprove Lamarckism by experiment and cut off the tails of rats for six generations and carefully measured the tail lengths of the newborn and the adults at the age at which he performed amputations. There was no shortening of the tail over generations. He then pointed out that an even longer experiment was in progress among Jews for almost three millennia. Jews circumcise their baby boys on the eighth day after birth, but there is no difference at birth in foreskin size between Jewish boys and non-Jewish boys (at four generations per century, this would be 40 per millennium or about 120 generations of continuous practice among Jews). A similar lack of effect was found for foot binding among Chinese women and lip size among Ubangi women, who stretched their lips with successively larger metal discs. He attributed tailless cats (such as Mainx cats) to new variations (in those days called sports or saltations and today called sporadic mutations) rather than to accidents (such as the wheels of wagons or carts cutting off tails) leading to an acquired heredity. Weismann ridiculed theories of maternal impression. Many mothers claimed that their sons or daughters had red marks on their face or other parts of the body because the mothers had witnessed, while pregnant, a burning building or other traumatic events. Such views are as old as the biblical tale of Jacob, working for Laban, who used visual stimuli (painted rods) to produce spotted progeny from the rams and ewes engaged in intercourse.

Weismann made an additional contribution as he considered the implications of the formation of eggs for his germplasm theory. Eggs produce what was called a polar body. Weismann argued that this was a means to reduce the genetic material of the germplasm by half. He predicted a similar reduction in genetic material in the production of mature sperm, which he termed "reduction division" in contrast to the simple longitudinal splitting of the chromosomes observed during ordinary cell division, which he called an "equation division". When meiosis was worked out in the late 1890s and early twentieth century, Weismann's prediction of a meiosis that leads to a reduction division was confirmed. But it was the full set of chromosomes found in the immature germplasm that was reduced by half. Weismann had accepted the chromosomal nature of hereditary material around 1885. We still accept the overall validity of Weismann's continuity of the germplasm and the separation of the germinal and somatic tissues during adult life. Attempts to bring back Lamarckism in the twentieth century have been discredited as consequences of faulty experiments, conscious or unconscious fraud, or politically forced belief. Despite the repudiation that Lamarckism has received, the belief in the inheritance of acquired characteristics is still widely believed by the general public.

Its most legitimate reincarnation comes from the study of RNA molecules acting as epigenetic factors that increase or decrease gene function, and sometimes these epigenetic factors become incorporated into the DNA of germ cells and thus act as a form of Lamarckian inheritance.

## Some Scientists are Good at Both Theory and Experimentation

Muller was both a great experimentalist and great theoretician. His major theory was the gene as the basis of life. Long before others accepted genes as central to all of biology, Muller pushed for this idea. He reasoned that genes have the unique property of copying their variations and thus whatever it was that gave genes a chemical or physical structure, that feature prevented errors from interfering with the copying capacity of genes. Without the ability to copy errors, evolution would not be possible and life could not perpetuate itself. Muller was particularly adept at squeezing a lot of theory from a few facts. He predicted that chemical mutagens would be found because he calculated that the amount of background radiation was insufficient to account for more than a minor fraction of spontaneously arising mutations. He knew from his own experiments that a ten degree rise in temperature doubled or tripled the spontaneous mutation frequency, which is consistent with the effects that chemists have found for temperature on the rate of chemical reactions. S. P. Ray Chaudhury, one of his students while he was at Edinburgh, found that the total dose of radiation received rather than how long it took to receive the dose determined the number of mutations induced in fruit flies. The rapid dose (about 30 minutes) gave the same mutation frequency as the protracted dose (about 30 days). To Muller, this meant that diagnostic doses also produced mutations because a 30-day stringing out of, say, 400 roentgens amounts to a rate of 9.26 milliroentgens [mr] per minute or 0.16 mr per second. A chest x-ray is about 0.1 r to 0.01 r (100 to 10 milliroentgens in an exposure that takes less than a second). This finding also meant that the public and their practitioners needed to protect themselves from radiation damage, but it was bitterly resisted for decades when he reported it in the early 1940s. Radiation genetics is still a controversial field because conflicts of interest, wishful thinking, and lawsuits have intimidating effects when science and politics or science and commerce are at odds.

## Some Scientists Extract More Insights From Their Data Than Do Others

Theoreticians can see not only the big picture but also the tiniest details in ways that escape most scientists. When Bridges and Muller independently discovered in 1936 that the bar-eye mutation was a duplication, both knew that they had an explanation for its peculiar high reverse mutation rate. Bar-eyed stocks kept producing normal round-eyed males at a rate of one every few thousand. But Muller recognized something that Bridges missed. With duplication, there is opportunity for gene evolution as each duplicated gene can mutate independently of the other over thousands of generations to come. From this Muller extended a descending set of laws: life comes from preexisting life (proven by Pasteur's experiments that ruled out spontaneous generation of life from non-life), cells come from preexisting cells (claimed by Rudolf Virchow (1821–1902 and Robert Remak (1815–1865) as a basis for cell theory and the single cell origin of cancers and other tumors), and genes come from preexisting genes except for the first gene from which all life on earth descended. He connected the implications of the bar duplication to the views he developed much earlier, on broad theoretical grounds, that the gene was the basis of all life on earth.

I have marveled many times when attending meetings and listening to scientists present their work or discuss their implications how these brilliant theoretical insights spontaneously come them. How they see it is by no means clear to me (or even to psychologists who study creativity). I suspect that artists, poets, composers, and other talented individuals also see these things in their own fields. Tracy Sonneborn, Muller's colleague at Indiana, told us with pleasure during one of his courses that at one meeting he attended in the 1940s, Muller listened to Avery, MacLeod, and McCarty report their findings on DNA as a transforming substance that shifted bacteria from one characteristic to another. Muller suggested that perhaps bits of DNA enter the cell and one of them pairs with its homologous region and undergoes crossing over to insert a genetic presence in the host bacterium. It may well be that Avery, MacLeod, and McCarty had also thought of this possibility but were too cautious to present it as the explanation without some sort of experimental evidence for it. What this example illustrates to me is not priority for an idea but the boldness and quickness with which theoretical insights arise.

## Experiments Frequently Modify or Lead to New Theories

I had a similar experience to Sonneborn when I heard Crick lecture in his early visits to the United States. Crick was struggling with the coding problem. He wrote a brilliant paper, *On Protein Synthesis*, in 1957 which presented his beliefs that DNA was the genetic material and used its sequence of nucleotides to specify the sequence of amino acids in proteins. To do this, a third component — a copy of the gene that left the nucleus and entered the cytoplasm — was needed. Many cell biologists during the 1950s reported that this suspected material was RNA and that it synthesizes in the nucleus and enters what is now called the endoplasmic reticulum in the cytoplasm. On theoretical grounds, Crick and other molecular bio-logists called this material messenger RNA (m-RNA). Crick predicted that something else was needed to translate or decode the messenger RNA. More specifically, there had to be "adapter" RNA molecules that recog-nized sequences of three nucleotides to specify amino acids. The triplet nature of the code was actually inferred on theoretical grounds even earlier (in 1954) by astronomer George Gamow, who reasoned that four nucleo-tides could specify four things, but a sequence of two nucleotides could specify $4 \times 4$ or 16 things. Since there are 20 amino acids that compose all our proteins, Gamow argued that the code had to be a sequence of three nucleotides, or a triplet (later called a codon), which would encode $4 \times 4 \times 4 = 64$ triplets. The predicted adapter RNA molecules were soon found and named transfer-RNA (t-RNA).

In 1959 at Brookhaven National Laboratories, Crick reconsidered the coding problem. He tried to work out a code that gave exactly 20 sense-making combinations out of the 64 possible triplet combinations, but his model was unworkable and he knew it. Theory alone could not generate a rational code. He reviewed all the evidence to support his belief that there was a genetic code and why a straight genetic approach by studying mutations would not yield the solution, eventually concluding that a new approach was needed. The following year while he was in Moscow at an international symposium on biochemistry, he heard Marshall Nirenberg (1927–2010) of the US National Institutes of Health describe his direct chemical experiments that used synthetic RNA molecules to solve the genetic code. Crick was alleged to have jumped up with excitement when he heard it, like King George III on hearing the hallelujah chorus of Handel's *Messiah* for the first time.

I like the retelling of the solution to the genetic code because it was stated first as a speculative theory (the "code script") in Schrödinger's *What is Life?* before being forced onto DNA by the ease with which aperiodic sequences could align themselves along the length of genes as double helix DNA. It was given its first new feature, the triplet nature of the code, on rational theoretical grounds by George Gamow. As experimentalists found the basis for what Crick called the "genetic dogma" (because it was initially taken on faith) that DNA produces RNA which produces protein, Crick speculated how this sequence of events could take place. Experimentalists soon detected the predicted messenger RNA and the predicted transfer RNA. Although theory failed to yield a correct genetic code, it did stimulate another experimentalist, Nirenberg, to use work that he and other experimentalists had amassed in the late 1950s to design an experiment that gave exact answers, such as the m-RNA triplet UUU encoding the amino acid phenylalanine — a finding that could not have been predicted then on theoretical grounds.

## Theory and Experimentation are Both Necessary in the Sciences

What this shows is that the two aspects of science — theory and experimentation — are intimately connected and need constant reinforcing. I am often puzzled by the simplistic way in which science is interpreted by non-scientists as if theory alone can rearrange things and make sense of some piece of the universe. It makes sense if it leads to experiments that churn out supporting or contradicting information from which new insights and connections are stimulated. It also makes sense if new data appears from new observations and these too support or contradict existing theories and force them to evolve. When we looked at Darwin's approach as a theoretician, we saw — if we are to believe him (and I do) — that it took some six years since he left on the *Beagle* before he had a theory emerge from an immense saturation of data of his own gleaning. It then took him some 20 additional years of reading, going to exhibits, corresponding with breeders, and reflecting during his daily walks through the gardens on his estate before he could piece together a story that he was convinced wouldn't be seen as slapdash. It was not theory and experimentation but theory and new data that gave Darwin the stimulation and feedback to construct his theory of evolution by natural selection. That new data came from exploration and discovery more than it came from experimentation, which was still relatively modest in the life sciences in the mid-nineteenth century. This example again illustrates that science is complex and cannot be reduced to a few simple procedures or ideas. The only thing that unites all that we have examined so far is faith in reason to make sense of complexity and the underlying reality that scientists try to study.

## Scientific Revolutions have Different Ways of Arising

Sometimes great theories are called scientific revolutions, and sometimes they do replace old ways of seeing or interpreting things. Sometimes they do not; they are just new and the novelty is the revolution. Muller's use of radiation to induce mutations revolutionized genetics not because it overthrew anything but rather added a new way to do genetics. When stain technology was developed in the late 1850s, making it possible to slice, mount, stain, and preserve specimens of plant and animal tissue and leading to the fields of histology and cytology, a revolution occurred in medicine and the life sciences, but nothing was overthrown. There was no way to study cells without microscopy. When Schrödinger conceived of a code script, he wasn't overthrowing some older way of thinking about how genes worked. There was no older way because no one had any idea how they worked. Schrödinger's gift of seeing genes as coding units was a great stimulus, but thinking of it in the confrontational sense that revolution traditionally brings to mind would be misleading.

A real revolution does exist when contending theories collide. When Kuhn wrote *The Structure of Scientific Revolutions*, he was correct in saying that the earth-centered universe of Dante's era (known to scientists as the Ptolemaic or geocentric model) collided with the sun-centered universe of Copernicus (the Copernican model). Galileo tilted the evidence in favor of Copernicus with his use of the newly discovered Dutch telescope, which he heard about from a correspondent. The rest is history. He saw the four larger moons around Jupiter. He calculated their orbits. He saw the crescent phases of Venus. The planets now differ from the stars. The planets are not perfect spheres. The planets reflect the sun's light; they do not generate their own. He saw the shadows cast by mountainous craters on the moon. The moon is made of rocks; it is not an ethereal body. He paid a price for merging theory with discovery and new observations. He shifted to describing the underlying reality and abandoned the approach to the universe based on inference from biblical scripture, which implied an earth-centered model of some sort. The scientific revolution more accurately contrasted one between Galileo's evidence-based model with Ptolemy's model of circles and epicycles, which was geared to a common sense and philosophically idealized assumption of earth centrality. The religious revolution, for that was what embroiled Galileo, was the rejection of a Scripture-supported earth-centrality in favor of a science and technology-supported sun centrality. In the sixteenth century, such clear distinctions may not have been raised in the court of religious law or the court of scientific reasoning. Even if they had, Galileo would have had to hedge his interpretations to appease the strongly religious intellectual world of his time.

In a curious way, Darwin's theory of evolution by natural selection involves a conflict not between two scientific doctrines (i.e., non-supernatural theories) on the origin of life but between an evidence-based theory of how life evolved (Darwin's) and a religious view, inspired in the same Scriptural source that generated the earth-centered universe and which implies the unique creation of species appearing all at once then as they appear now and fixed in kind (the Bible uses "kind" and not species because the species is a Linnaean concept).

A true revolution (in the contentious sense) is Weismann's theory of continuity of the germplasm with its rejection of the theory of acquired characteristics. In the contest between Weismann and Lamarck, Weismann won. Virtually all the experimental evidence for the non-somatic origins of germinal mutations favors Weismann's model. Most of what presently passes as scientific revolutions in the life sciences — cell theory, the theory of the gene, and molecular biology — are additions to knowledge rather than an overthrowing of old and contending scientific views.

I urge those readers favoring the idea of paradigm shifts as the basis for scientific progress and change to read my book, *How Scientific Progress Occurs: Incrementalism and the Life Sciences*. I provide a detailed argument favoring incrementalism as the major way the life sciences have evolved.

# 7 Science as Technology

Scientists frequently use tools. Tools are frequently extensions of our five senses. They can do tasks better and transcend the limitations of our evolutionary gifts. We can see objects that are as small as a speck with proper contrast, like peppered dust on white paper. We cannot see our cells. We can squint as we might but we cannot see Jupiter's moons or Venus's crescents, but we are more aware of planets as bigger objects than a point because they keep changing in size when we observe them seasonally with our naked eyes. Stars are relatively constant. It is not surprising that the first scientific tools are means of enlarging the familiar, such as telescopes and microscopes. I am still thrilled when I use a telescope to look at the planets or resolve the Pleiades into a cluster of stars. I have not yet had the experience of using a professional telescope to see a nearby galaxy like Andromeda. Someday, I hope that opportunity will arise. Most of the wonderful objects in space that I have admired in book plates or magazine articles are not seen by a glimpse through an eyepiece. They are slowly recorded by photographic film or photon sensitive devices that are enhanced by computer programs to produce faint portraits of faraway galaxies no eye can ever visualize.

## How Galileo Changed Science Using a Telescope

The first telescope was constructed from two lenses and a tube in Holland in 1608. Exactly who gets the credit is still in dispute, but the three contenders are Hans Lippershey, Zaccharias Jansen, and James Metius. Tradition favors Lippershey's version — he was a spectacle maker and he happened to look at a church steeple using two lenses instead of one, and as he held them apart and resolved the blurred image, he was startled to find it enlarged. He took mental note of the approximate distance, did a more precise measurement of focal length, and then mounted the lenses in a tube. It is a believable tale and I can imagine other spectacle makers of that age trying the same thing and, after a moment's amusement, putting the lenses back in their display cases and thinking nothing more about it. The inventor has to add some intentionality to the discovery. Spectacles came into use during the Middle Ages, which is a good three centuries or more after lens making became a craft. Why did it take so long? I suspect many a discovery of the telescope principle (synergistic enlargement by use of two lenses) was made over those years but the application to construct a telescope was lacking. Spectacles are for people with poor vision. They are not toys. Lens makers were healers, not warriors, and they may not have thought of military applications. They may also have had no interest at all in astronomy. We can infer that from Lippershey's own failure to apply his telescope to the skies and record what he saw. Galileo heard about the instrument at a meeting in Venice and made his own telescope. It was a crude instrument with blurry images (3 power by today's measurement), and he had to lie flat on his belly and prop the telescope on his window sill to use it. He became adept at polishing lenses and making more refined instruments, his best being 33 power, which was the telescope that gave Galileo his startling discoveries that there was a real world out there that science could describe. It was no longer homogeneous light from the skies. There was reflected light from the planets that had to be from the sun and there was light too far away — from the stars — that couldn't be resolved into planetary bodies.

## Invention Requires Imagination for New Uses

Galileo's awesome accomplishments with his telescopes make me aware of how important it is to combine the instrument and the user for maximum exploitation of an invention. Lippershey may well have invented the telescope, but he did not conceive of using the telescope for astronomy. Perhaps, like Galileo, his first instrument only magnified 3 power, which would be good for church steeples but not for resolving Jupiter's moons or Venus's crescents. Spectacle makers were not making lenses that magnified much more than that, and if Lippershey had no scientific interests, he would not have sought to magnify more than what was useful around town or in the meadows. What applies to inventions also applies to ideas or theories. A few persons claimed they had the forerunner idea of natural selection before Darwin. They may well have conceived it, but they did not follow up on it and their published note was just a speculation soon forgotten. It is not uncommon at scientific meetings to hear about works in progress and an idea picked up suddenly ignites and shoots off in new directions. The person who heard the rumored discovery or offhand comment may no longer remember the source when the project is completed and ready for publication.

Although Lippershey gets the credit for the telescope, Jansen is assigned priority for inventing the compound microscope, leaving poor Metius a perpetual contender. The discovery was even earlier, in 1590. Good microscopy was first reported, nearly a century later, with a compound microscope by Robert Hooke in England and Marcello Malpighi in Italy. Hooke gave us the idea of cells (1665) and Malpighi discovered the capillaries (1668) that connected the blood supply from the arteries to the veins. Just as remarkable were the observations made with a simple microscope (just a fancy magnifying glass, really) by Anthony van Leeuwenhoek in the 1680s.

## Hooke's Microscope Opens up a New World of Life

Hooke published the *Micrographia* in 1665. I held the first edition one summer afternoon in the vault of Indiana University's Lilly Library, which houses one of the finest collections of rare books and manuscripts in the world. I was working on Muller's biography and classifying his papers as I was using them, and I had been given librarian's privileges and my office was in the vault. One day a batch of books appeared in preparation for an exhibit. I was suddenly alone with the most famous books in the history of science. It was as close to apotheosis as a scholar can hope to experience! The *Micrographia* is a hefty book, not the sort of thing you can nicely prop on your chest and read in bed. Its thick white paper is as crisp today as it was when printed more than 300 years ago. The engravings of Hooke's insects, sections of leaves, and his famous piece of cork are gorgeous in their detail. Hooke did not conceive of a cell theory, but he did see cork as composed of cells. His cells were empty boxes and they immediately suggested to him why cork was so buoyant. He called them cells from their cubical shape, and they also looked to him like cells used by monks. I often wondered about this image until I was once in Stockholm visiting my father's birthplace (the home of Carl Bellman, the Swedish equivalent of poet Robert Burns). Across the street was an old Church, St. Margaret's, built in the thirteenth century and remodeled in the sixteenth century. Along the courtyard were small salmon-pink cubical dwellings. These had been used by monks until King Gustavus Vasa adopted Lutheranism as the state religion. They were now used by Greek Orthodox monks.

Hooke described the thin slice of cork he placed on a black background as resembling a honeycomb. He also likened the pores or cells to "little boxes" and attributed to them the explanation for the buoyancy of cork. Hooke did not have a sense of the community of cells we associate with cell theory. That came much later, in 1838, when Matthew Schleiden and Theodore Schwann proposed such a model for all animals and plants. Their cell theory was flawed as they believed that cells crystallized out of fluid, mistaking nuclei as early condensates of a cell. Their form of cell theory is sometimes called "free formation of cells." Curiously, their theory was revived during the Cold War when Soviet attacks on Western science even included cell theory. One Russian biologist, Olga Lepeshinskaya, claimed that she had observed the free formation of cells from albumin. Despite their flawed model of the origin of cells, their essential belief remained unchallenged — plants and animals are composed of cells, and it is the contents of the cell, not the boxes as containers, that compose the cellularity that biologists had to study.

## Ideas and Theories Evolve in Science

Note that this discussion of cell theory shows something that most historians of science appreciate, the public does not know, and most students hate. Ideas and concepts evolve as they get modified by new observations and, on occasion, experiments, to reveal a detail of structure and function not even imagined by Hooke. I remember one graduate student from the dormitory where I resided during my first year at Indiana University who despised being taught all those theories that were proved wrong in his physiology class. "Why don't they just teach us what's true and forget about all this history?" He lamented. It is a wonderfully simple but false idea of reality — that there is a truth which, once found, remains constant. The duty of the teacher, he felt, was to gather all the nuggets of truth and present them. It assumes that there is no future modification and new knowledge doesn't change the old, it only adds brand new nuggets.

Microscopes, as we saw, were in use since the 1590s but did not get a splash of publicity for science until the mid and late 1600s. At that time, Leeuwenhoek reported seeing bacteria, spermatozoa (from dogs), protozoa (in his own mouth), red blood corpuscles, and the stages of the life cycle of weevils and other insects. He revealed a world of minute organisms, like rotifers, that existed wherever stagnant water could be sampled, and the earth is teeming with abundant life, each one with its own life cycle. He published regularly in the *Proceedings of the Royal Society* and was elected to membership by that society. He and Hooke inspired many people to buy microscopes (including diarist Samuel Pepys), but no major advances in microbiology or cell biology came until the nineteenth century.

## Scientific Progress Follows an Erratic Path

Long gaps rather than steady progress typify the way science works, and we puzzle over the reasons for these delays. In microscopy, much of the blame is associated with the twin problems of lens aberrations. Some lenses cause light rays to break up into concentric rings, making the object look fuzzy. This phenomenon is known as spherical aberration. Other lenses break up white light into rainbow patterns and different colored images are seen somewhat out of focus, which is called chromatic aberration. For telescopes and microscopes to be good, a solution to these problems was required. Several scientists came up with the solution, including Joseph Lister's father who developed lenses composed of two different glasses, one containing lead and the other without dissolved lead. By altering the density of glass and using a fused sandwich of the two different glasses, Lister obtained a single lens that was corrected for both its optical and spherical aberration. Good compound microscopes began appearing in the early 1800s and these allowed for more accurate observations of cells, including the discovery of the cell nucleus by Robert Brown.

Until the 1850s, slices or smears of tissue were observed by microscopists who had to be ingenious in finding good contrast and oblique lighting to reveal structures within cells and the shapes of cells in different tissues. A major advance in technique came through the development of stain technology. As microscopes improved in the magnification and clarity of their images, many European physicians began to examine normal and pathological tissue. Schwann's treatise, *Microscopical Researches*, was the first extensive study of all the tissues of the human body as well as samplings from many other animals. He confirmed Schleiden's findings that all plants were composed of cells and extended it to animals. His treatise covers sections on isolated cells (blood), continuous tissues (epithelium, the eye lens), cells with matrices (cartilage, bone), fibrous tissues (fat, tendon, connective tissue), and muscles, nerves, and capillaries. He could see within the nucleus and named the globular object he saw a nucleolus, mistakenly believing it to be a little nucleus that would grow and form a nucleus. The nucleus growing and forming a cell as part of the free formation of cells was the explanation that he and Schleiden offered for how new cells were produced. Mitosis and its relation to cell division would not be developed for another 40 years.

## Stain Technology Developed in the Mid Nineteenth Century

One microscopist studying tissue anatomy, Joseph Gerlach (1820–1896), attempted to stain tissue slices with dyes. In the 1850s, most dyes were natural products like cochineal from the mealy bug or indigo from the root of woad. Steeping tissue in dye for a few minutes was not effective and Gerlach, not a very tidy person, left some of his specimens on his table without cleaning up. When he returned the next day and looked at the tissues before discarding them, he found to his surprise that the color had taken and he could see very clearly the contrast between the nucleus and the surrounding cytoplasm. Gerlach and other microscopists then developed the techniques for preparing slides. When I was an undergraduate, I took a course in histological techniques which I thought might come in handy someday. I learned that quite a bit of work went into making slides. I also liked it because the laboratory reeked of the odors of alcohol, xylene, and paraffin. I began by taking some tissue, say a piece of gut from a mouse, which would be dropped into a solution that preserved it without shrinking or distorting the tissue too much. It might be formaldehyde or an acid (picric acid was a favorite). The tissue was then removed and put in alcohol until all the water was extracted, after which it was put in a solvent (xylene) that could be penetrated by melted paraffin. Once the paraffin hardened, the wax was cut and shaped until the tissue looked like a bit of headcheese. The paraffin was glued onto a cutting instrument called a microtome, and the blade (like a miniature baloney slicer) produced ribbons of cut sections of the specimen. These ribbons were next mounted on slides using egg albumin as a glue and dipped into jars with different solvents. The first would contain xylene and leach out the paraffin, followed by 100% alcohol to leach out the xylene. Then a series of jars would contain stains dissolved in alcohol, such as eosin or hemotoxylin. As each step was successfully achieved, the slide would move from one jar to the next (the jars are called Coplin jars). Finally, the slide would be placed on a drying table or hot plate and some resinous material would melt over the surface. A thin glass cover slip would be lowered on to it, carefully avoiding bubble formation, and the slide would then be labeled with a small slip of paper on one side describing the specimen and the stain. Each step involved the work of a different scientist, and each step was trial and error to figure out how long to steep a slide in the Coplin jar. Each dye was tried out by an investigator to see how good it was and what it stained in the cell. There were hundreds of scientists in Europe doing this in the late 1850s and early 1860s when much of the technique was developed.

## Sometimes Luck Enables a Science to Advance

I have gone into detail to give some idea of the lengthy process involved in making quality slides. Pathology laboratories still use these techniques to prepare slides for examination. Biologists still use it to describe new plants and animals and get an idea of the developmental processes taking place. One of the pleasures I had in an embryology course as a graduate student was reconstructing in my mind the three dimensional portrait of a pig embryo from the one hundred or more serial slices of the embryo cut from head to tail. It is a very labor intensive process and takes one or more days to make good slides, but no doubt today much of this is automated for hospital laboratories where many specimens come by each day. This story is worth telling because it illustrates a feature that is often downplayed because it is not very rational — science is often advanced by luck. If Gerlach was a neat guy, he would not have discovered the combined effect of heat and time needed to get the dye to set in his tissues. He had left them on a warm coal stove and was probably too busy or tired or it was simply late and he had to go home without cleaning up. In scientific publications, one doesn't confess to being a slob or being lucky. Papers merely present the techniques and not the circumstances that lead to the finding of those techniques. Published articles are mentally tidied up and thus we infer, falsely, that the same tidy process must have gone into the discovery or invention. We also infer, falsely, that scientists are faceless people, all brain and reason with no emotion or humanity. A very similar untidy process goes on in artistic and literary creation, but the nature of the product is packed with feeling and emotion, although not necessarily the emotions and circumstances that led to the creative work. We do not think of writers as just sitting down and writing a chapter a day without corrections, breaks, discarded sheets of crumpled paper, and occasional fits of disgust or frustration.

There is one last piece of the stain technology story that is worth mentioning. William Perkin (1838–1907) was a young English student with a talent for chemistry and was sent to study at the Royal College of Chemistry in London. He was just 18 years old when his instructor mentioned that it might be possible to synthesize quinine or quinine-like products. He tried his hand at it, often working at home in a crude laboratory that he had set up. He discovered a blackish substance which, when dissolved in alcohol, produced a gorgeous blue to purple dye. He called it mauve and it was the first of several aniline dyes he synthesized. His synthesis of alizarin, a brilliant red dye, replaced both madder and cochineal as the dye of choice for staining fabrics red. He got his father and brother to help him exploit the dyes he patented, but for some reason the English were not as quick to apply the dyes to microscopy. This application was done in Germany and the aniline dye industry that Perkin started in England was soon eclipsed by its commercial success in Germany. Perkin was one of the youngest scientists to make a major contribution to knowledge. The only thing comparable in my mind is Sturtevant's discovery of mapping chromosomes from the crossover frequencies of their genes when he was Perkin's age.

## Laboratory Work Leads to the Invention of Tools to Make Work Easier or Possible

Scientists sometimes invent tools to do their work, often without patenting them or seeing commercial value in them. Sometimes they publish the item as a brief note, as did hundreds of fruit fly workers in an informal publication called *Drosophila Information Service* (abbreviated and called *D.I.S.*). Thus, when Morgan first started to use fruit flies, he used any type of glass container he could get, often a pint milk or cream bottle. He mashed bananas against the bottom of the jar and stuffed tissue paper at one end to form a dry pad for the flies to alight. He bought the bananas out of pocket, and he and his students brought in their empty bottles (or swiped some from stoops on their way to Schermerhorn Hall). Morgan would use a jeweler's loupe (an instrument of brass that one can slip in one's pocket with two lenses close together) to look at the flies. If he saw a fly he was interested in, he would take an empty bottle, remove the plug of cotton, and put the bottles neck to neck. He would separate and plug them and see which bottle contained the fly he desired. He would then cut the contents in half again until he finally isolated the desired fly. His student, Calvin Bridges, thought this was tedious. Morgan also did his work standing up which made it tiring. So Bridges set to work and introduced the dissecting microscope, a porcelain white plate on which to put flies, and a small bottle with a funnel that terminated in a perforated gelatin capsule, which he called an etherizer. By knocking out the flies and moving them about with a watercolor brush, Bridges greatly increased the ease of doing fly work. The flies could be sexed, counted, and sorted into desired character traits. Bridges also standardized the making of food for the flies (scientists like to call the food "media"), although not without occasional disasters such as fermentation explosions of overripe banana mash. There are hundreds of labor-saving and clever devices for working with fruit flies in *D.I.S.* Scientists constantly tinker with the equipment they use and find clever ways to make them more useful. Usually they do not lead to new techniques or instruments, but they do make the work faster or easier to do. Quite a few of these improvised inventions are unique to the laboratory and never written up. In my early days as a geneticist, I used to lecture on my work in different universities and I would get a tour of the fruit fly laboratories. No two laboratories had the same design for an etherizer, and I was always intrigued by some of the devices I saw. I retained in my study one that was used by Muller.

# 8 Science and Values

Scientists like to think of themselves as objective, and the public often shares that belief. In this shared belief, the scientist will set aside personal biases and pursue the findings with a "let the chips fall where they may" attitude. There are two assumptions that are worth exploring. Is knowledge itself neutral? Can a scientist ever be free of bias? I like to think of the neutrality of knowledge by citing two examples. I sport above my left knee a small bluish dot that serves as a reminder of sibling rivalry. About 80 years ago, I was pestering my brother and he took a pencil and stabbed me in my leg. I was about 7 or 8 and he was almost two years older. I do not know who first invented the pencil encased in wood, and I doubt that that person considered anyone would be using the pencil as a stabbing instrument. I would not expect that inventor to have put effort thinking about how the pencil could be abused. My simple wound is minor compared to the ransom notes, plans for bank robberies, murder plots, and pretexts for wars that were first jotted on a slip of paper with pencil. Is the pencil inherently evil? Of course not. Is it value neutral? Only in a naïve sense that the pencil was not designed to be used for evil intents. Its uses for writing poetry and literature, composing music, writing up plans for inventions, drawing landscapes and portraits, designing a stage for theater production, and planning and recording experiments far outweigh its uses for nasty purposes.

## Devices Designed for Military Use are not Value Neutral

Now imagine a person hired as an engineer by a defense contractor. Let us say the engineer is an inventor and the employer is a manufacturer of guns and other small arms. The inventor is asked to design a gun that can fire 10 to 20 rounds of bullets per second and hold about 100 or more bullets. The inventor knows that the company receives most of its funding from the government and that its biggest purchaser is the military. The engineer does the job and a beautiful semi-automatic weapon is the outcome. Suppose the engineer is asked about his role in bringing about the deaths of several tens of thousands of soldiers killed in skirmishes by those using this weapon. Or, with more horror, the deaths of children in schools, shoppers in Walmart, parishioners in churches and synagogues, or teens enjoying a dance at a night club caused by a crazed or misguided zealot using that semi-automatic weapon. Could he honestly reply, "My job was to solve a scientific problem. How my design was used was not a matter for me to consider because solving problems is value neutral. Only the users of the gun, not the designer, bear moral responsibility." This is an argument often made by those who design instruments that are used for evil purposes. The chemist who recommends the amount of cyanide needed for gas chambers, the engineer who designs a cremating chamber that can hold several dozen bodies and use the melting bodies as fuel, or the architect who lays out the plans for placing incinerators, fake showers that spray cyanide gas, and temporary facilities to marshal concentration camp victims before their destruction may claim neutrality, but post-war trials of Nazi scientists often convicted these people of war crimes. It is difficult to believe that one can design a machine gun and not know how it will be used.

## Scientists May not Foresee Possible Bad Outcomes from Their Work

Not everything falls neatly into innocent inventions and willfully designed instruments of death, torture, and destruction, which makes it more difficult to assess blame if wrong outcomes result from scientific and technological activities. A good example is the field of radiation in physics, medicine, industry, and the military. In all four fields, unhappy things arose. But one could successfully argue that despite these sad outcomes, our knowledge of the world was enriched — medical uses of radiation have saved millions of lives, billions of dollars and thousands of lives have been spared industrial accidents, and a war was ended faster than it might have, sparing many hundreds of thousands of lives had a lengthy invasion of Japan been the price of a non-nuclear end to World War II. All of those examples bring into play an operational value system called utilitarian ethics that weighs good versus bad and makes a quantitative estimate that the good outweighs the bad. Religious values are often based on absolute right or wrong — murder is never justified; theft is never justified; lying is never justified. Sometimes these are inferred universal values based on reason and not religious belief, as philosopher Immanuel Kant argued. Reason would lead us to these absolute values.

## Both Utilitarian and Universal Values are Used by Scientists

People are too complex to be consistent. We are sometimes utilitarian and other times Kantian depending on the issues and principles at stake. Sometimes we are not ethical at all but just self-serving. We cannot help but be contradictory because that is what distinguishes human behavior from the behavior of saints, whose unusual enlightenment and commitment make them consistently moral. This is also true for scientists. When Wilhelm Roentgen discovered x-rays, he was quickly aware that his technique could be used to detect metal fragments embedded in flesh from accidents and gunshot wounds. He had no idea that x-rays would also be used as a treatment for cancers, which came about as hundreds of physicists and physicians used x-rays to study their uses. Some of these early practitioners noted a reddening of the skin from exposure to radiation. Unlike sunburn, it was slow to heal. Sometimes ulcers arose that did not heal and within about five years a few practitioners developed cancer in their fingers or other exposed parts of their bodies. Several dozen died in those early years from radiation-induced damage. Thousands more had reduced life expectancies as they plied their radiation diagnoses and treatments over the decades and took scant interest in radiation protection. Until 1927, no one knew that x-rays induce gene mutations or chromosome breaks. Until 1945, the condition called radiation sickness was largely unknown to both the general public and specialists alike.

X-rays quickly took hold in society. Physicians learned that they could diagnose ulcers, tubercular lesions, and the tell-tale presence of primary and metastatic tumors. They could even see where the baby's head was when it came time to give birth. They learned that radiation in high doses could destroy tissue, especially that containing dividing cells. Some even thought x-rays had a stimulatory effect and it was used to treat arthritis or glands associated with fertility, such as the pituitary and the ovaries. Some used high doses to induce temporary sterility in males by zapping the testes. Some straightened out bow legs in children by irradiating the more rapidly growing arc of the long bones. Those who did these treatments in the 1920s had no worries about bad outcomes except for the troubling knowledge that occasional notices appeared in medical journals about colleagues who had died from radiation-induced cancers.

## Radiation Protection is Inherently Controversial

Warnings for radiation protection were largely non-existent and precautions were modest. X-ray units were sold to shoe stores with the advertising claim that only x-rayed feet could provide an accurate fitting of shoes and save the customer from corns, bunions, and cramped ugly toes. I saw one of those units in 1958 when I taught in Kingston, Ontario. The proprietor no longer used it and wanted to know if I was interested in taking it and using it at the university. I told him I was more interested in how many children he examined to fit shoes so that I could estimate the number of leukemias he might have induced. He was happy to see me leave. I knew it wasn't his fault — he had been duped, the way so many people are duped, into using a product that they really didn't need. It was just a gimmick to sell more x-ray units by a corporation that had filled virtually all examining offices in hospitals, dental practices, and private health practices, from MDs to chiropractors. They had no idea back then, in the 1920s, that there were serious hazards ahead. They probably did know that a disproportionately high number of physicians and physicists who pioneered radiation technology were dying. One major corporation that manufactured x-ray units contacted T. H. Morgan and asked him if he was interested in studying radiation effects on biological materials. He wasn't keen, so they contacted a local biologist, John Mavor, in the early 1920s. He found vague effects such as increases in numbers of flies that had gained or lost a chromosome and changes in the distances between genes when studying x-rays and crossing over.

## Muller's Radiation Experiments Extended the Controversy on Radiation Protection

Muller's work on the induction of mutations by x-rays transformed the field. Radiation genetics became a major activity of biologists and hundreds of papers followed within a few years after Muller's publication of his startlingly high frequency of induced mutations. Despite the publicity over the link between x-ray exposure and mutations, many physicians continued to use x-rays to treat fertility well into the 1950s. The publicity over Hiroshima victims in books like Hersey's *Hiroshima* eventually turned the public against radiation, and the Cold War hastened this shift in values because the US lost its monopoly on nuclear weapons in the 1950s.

Because science permeates so much of society, it is difficult to separate scientific values from those we expect in the business world or any other aspect of life. Corporations and individual entrepreneurs may make decisions that are insensitive to the community, employees, or customers. They may establish unsafe working conditions, deliberately withhold information about hazards they know to be present, or produce defective goods or products that put the customer at risk. Sometimes the decision-makers are scientists themselves. Often they are not, and it is purely a coincidence that scientists work there, so they may or may not know of the abuses going on. When they do, they face a crisis of conscience that all potential whistle-blowers face. When do you speak out? Is this something worth losing one's job over or should one put livelihood, such as supporting one's family, over the remote risks others face?

## Distinguishing Risks from High and Low Doses is Sometimes Ignored by the Public

A group value system can also develop. Many physicists felt that concerns over radiation hazards, whether through fallout from weapons testing or potential accidents in the use of nuclear reactors to generate electricity, were irrational. Physicists could not understand how minuscule doses of chronic radiation in the daily operation of a nuclear plant could pose hazards. The yearly exposure was often less than the exposure that passengers experienced while flying from coast to coast on a jet flight. People who smoked one pack of cigarettes per day for 30 or more years were thousands of times more damaged by their habit than a worker exposed to operating doses in the plant for the same period of time. I do not doubt that this is true. Many people do not distinguish high and low risk. At the same time, there was enormous overconfidence in both the design of nuclear reactors and the ability of these machines to be "idiot proof" when workers, including skilled engineers, make mistakes. Both Three Mile Island and Chernobyl were preventable disasters and were not supposed to have happened. The nuclear engineers both here and in the USSR were convinced by the findings of the Rasmussen Report, which calculated the chance of a nuclear accident as one in 10,000 years. While this probability is higher than the chances of a Cretaceous-Tertiary impact meteor or comet hitting the earth and producing mass extinction (about once every 60 million years), it was intended to be reassuring. Such an event would be remote, at least for historical times. It turned out to be false. The Rasmussen Report, written by rational engineers, did not consider that engineers could be irrational and turn off safety devices as a short cut. They did not anticipate that in the event of accidents, people — even scientists — panic and may do the wrong thing.

Utilitarian ethics prevailed and in its modern form is sometimes called risk-benefit analysis. One calculates the chances of nuclear accidents (1 in 10,000 years) and then weighs that against the benefits, such as a pie in the sky early estimate that electricity would be so cheap it would virtually be given away free (touted to the public back then by shows such as *Our Friend the Atom*). When more realistic assessments of costs sank in, these were offset by the belief that the more damaging effects of combustion product hazards from coal and oil to the environment would be eliminated with clean nuclear energy. The problem with such calculations is that they contain hidden assumptions. If the accident was as serious as Chernobyl, what would have happened had it occurred at Three Mile Island? Would such a release of radiation across huge chunks of Pennsylvania have caused hundreds of billions of dollars in cleanup costs? No doubt it would have. The nuclear industry did not have to fear bankrupting lawsuits to pay for the health risks of those who might have been exposed to huge amounts of atmospheric radioactive elements. Congress had protected the new industry by limiting the awards in case of nuclear accidents.

## Is Basic Science Inherently Value Neutral?

Although applied science is by its very nature an application of values, to what extent is basic science value neutral? What types of research should be banned, if at all, because some future abuse might be more frightening than any present scientific abuse? Among such concerns were the development of recombinant DNA technology and the recurrent alarm over the cloning of human beings. Scientists who face these fears usually dismiss them as irrational fears of the public and comfort themselves with the belief that these are the imaginations of scientifically illiterate Luddites. There are some environmental groups that commit acts of violence to get their political message across. Such terrorist tactics are adopted by groups representing ideological, political, or religious values with mixed success.

Recombinant DNA arose from the discovery of a category of proteins called restriction enzymes, which recognize specific sequences of DNA and cut them only at those sequences. In an invading bacterium, there may be a half dozen or more of a particular sequence in the chromosome. If the body has an enzyme that can cleave the bacterial chromosome at these places, the bacterium will lose its capacity to divide. The same would be true of a virus even if it had only one such recognized sequence. Those who made this discovery realized that it could be used to open a virus or a synthetic small chromosome, and a segment of DNA from a different organism (say a human) could be inserted into the virus or synthetic chromosome (called a vector). Scientists recognized that there were few limits on how this technology could be used. Some beneficial uses include making human products, such as human insulin or human growth hormones, or life-saving enzymes that dissolve clots blocking a cardiac artery during a heart attack. They also saw that it could be used for germ warfare development, where ghastly toxin-producing genes might be inserted into common germs that are easily dispersed and inhaled. They also wondered if, in the process of doing something beneficial, an accidental insertion of a gene might render a once harmless bacterium or virus into a devastating pathogen.

To resolve their own fears, Paul Berg, one of the discoverers and developers of this new technology, got together with James Watson and drafted a letter to the top scientists who were likely to be using this technique for basic research or future applications. They held a conference and invited science reporters and staff from government science agencies to cover their sessions. I remember well the debates when they were discussed by my colleagues in molecular biology, some fearful of unanticipated hazards and others confident that such catastrophes cannot happen. For those who had no fears, there was a confidence similar to the physicists and engineers who were buoyed by the Rasmussen Report. It was a gut faith more than an actual study that convinced them. They felt that our genetic systems are manipulated by nature in all sorts of ways all the time and no epidemic ever destroyed all human life. Even the bubonic plague had only wiped out about one-third of Europe in mid-fourteenth century.

## Recombinant DNA Technology Began with a Self-policing Approach

Their colleagues who felt uneasy were not so sure. Recombinant DNA was a new technology and they wanted a go slow, one step at a time approach to launching the field. The scientists came up with their own solution at a conference in Asilomar, California, where a series of physical and biological containment procedures, including forbidding research until the state of knowledge could rule out imagined hazards, were recommended.

Recombinant DNA technology is now a flourishing field involving thousands of laboratories and many experiments have been carried out. The safety record has been excellent because most of the imagined catastrophes are biologically impossible and cannot occur. I do not doubt that the ideas can be applied to germ warfare, but the techniques to produce such pathogens are not dependent on recombinant DNA technology. Ordinary genetic means do exist that can lead to the creation of some pretty horrible germs. What protects us from their formation and potential use is their King Midas touch property. Once a pathogen is released, it can come back and devastate your own country. The movement of released pathogens cannot be easily contained.

## Higher Values are Invoked by Scientists Whose Work Might Lead to Bad Outcomes

I have thought about the people who do work on germ warfare and atomic weapons development. What are their values? Most of them, I have learned, are nice people you would enjoy as neighbors. This realization depresses me. I knew one colleague at UCLA who was friendly, a good teacher, loving to his family, and a Scout master. He also enjoyed playing tennis and had no rigid ideology guiding his politics. But every summer, he flew to Camp Dietrich in Maryland to participate in their germ warfare research program. He never talked about what he did, and I never asked. He no doubt had a Soviet counterpart who was equally affable and working with the same efficiency and dedication in their own germ warfare facility, and who probably shared the same values of patriotism and making the world safe through a strong defense. What is absent here? I believe that it is our reluctance to look at war and the justification of killing in war as a psychological disorder. We are used to thinking of war as economic, ideological, religious, or political in origin. A few psychiatrists, like Erich Fromm and Maurice Walsh, have argued for a psychiatric theory of war, but their efforts have largely been dismissed. Who in military command would like to think of themselves as a smartly dressed lunatic?

## Is Human Cloning a Red Herring?

With the cloning of the first mammal in 1997 by Ian Wilmut, a sheep from a healthy adult breast cell, a team of Scottish scientists revived a much discussed fear. Cloning of humans has always been a red herring for me. I find little likelihood in it ever being widely used, and there are many reasons for my dismissal of public fears. Clones are identical twins, and having known many identical twins, I know that they differ from one another in personality and behavior, sometimes quite profoundly. Clones can never be clones of the same person's psyche. While a person may get some narcissistic thrill out of seeing a younger version of herself or himself, it would only be a superficial resemblance. The values of the clonal child may differ substantially from the parent's, and the stress on the parent-child relationship might be especially frustrating because of conformity expectations on the part of the parent and loathing on the part of the child who sees his or her future self as seriously flawed. We can be more forgiving of our differences and imperfections when we see each other as genetically unique. I also believe that any drill sergeant can whip into shape one hundred 19 year-old males taken from any large campus and they will be as obedient, cohesive, synchronized, and similar looking in their smart uniforms as any clone of one hundred 19 years-olds derived from a single individual. Moreover, I doubt that the world's diversity of genes amongst 5 billion people will be affected if a few thousand relatively wealthy and idiosyncratic individuals choose to clone themselves. Most young parents are interested in sexual reproduction, not cloning.

When amniocentesis first appeared, there were fears that it would lead to an epidemic of aborted female embryos in the US and Canada. However, this never happened and hundreds of thousands of women have undergone amniocentesis on the recommendation of their physicians. When a sperm bank for geniuses using Nobel laureate sperm among the donors was opened in Escondido, California, some fifteen years or so ago, there were fears that people would widely use this material to produce genius children. This program received wide publicity and was written up in numerous national magazines. Although the sperm was free for the asking, only 250 children were born since the program began, which is hardly a tiny fraction of the artificial inseminations that occur each year for infertile males. That number is estimated at about 20,000 per year. When the founder of this germ bank for geniuses died, so did his sperm bank. The published results were disappointing because many of the users and donors cheated and lied while filling out questionnaires. None of the sperm came from actual Nobel laureates; theirs was too old. Cloning humans makes good science fiction but has low appeal for popular practice.

# **9** Science and Fraud

Every field of human endeavor is subject to fraud. We read of ministers who have been convicted of stealing the money given by their faithful parishioners and squandering it on personal luxuries. We read of legislators who put phantom employees on their payrolls and funnel the money to themselves or to their relatives. We know of innumerable instances of persons using bribes to get contracts, win government projects that utilize their companies, and have shoddy work practices approved. People sometimes lie about their education, experience, and intentions to get jobs and promotions. Such bad conduct also exists among scientists.

Scientific misconduct and fraud are a surprise to many people because scientists are looked upon as seekers of the truth. This, of course, could also be said of ministers, rabbis, priests, and other religious leaders, and we know that fraud exists among all religions. Dante portrayed several popes in hell for their wickedness, which included such corrupt practices as buying high church offices. Throughout the bible, we encounter examples of deception and fraud beginning with Adam and Eve themselves. Although society as a whole accepts these lapses of judgment and restraint as characteristically human behavior that is universal across cultures and generations, scientists frequently minimize the incidence and importance of fraud.

## Scientists Often Believe that Science is a Self-correcting Enterprise

Many of my colleagues believe that science is spared from a high incidence of fraud because science is self-correcting. If a person fabricates an experiment claiming that a certain process brings about a surprising outcome, the publication of that claim will prompt other scientists to attempt to replicate the work or use it to extend their own work. Let us assume that a physician fakes data to claim that colorectal cancer cells injected into rabbit blood produces an antibody that can be isolated and then injected into patients with colorectal cancer. Let us say the investigator claims that the tumors regress and the cure rate of colorectal cancer is 80%. There are tens of thousands of colorectal cancers every year in the United States. This news, if published, would be very exciting to physicians who deal with colorectal cancer patients. Perhaps ten or more laboratories would immediately try to repeat the work of the now very famous physician-scientist who made the claim. Suppose the work cannot be confirmed — what would scientists do to resolve the two different results? The scientists repeating the work will write to or call the first investigator to find out what they may have done wrong or differently. There will be an assumption of honesty because that is what scientists expect when they read a publication. They acknowledge that errors are common and thus either the original scientist made some sort of error or the repeating scientists made errors. If the repeating scientists are not satisfied, they will urge the original scientist to repeat the work and provide more detail over the procedures used. Failure to cooperate will lead to an immediate suspicion of fraud, and so will a sequence of delays for a variety of personal reasons. Sooner or later, the original scientist has to deliver the goods — a detailed account of the experiments with every possible step used so that the other investigators can find errors either of their own or of the original investigator's. They will carefully check the statistical significance of the data. Very frequently, invented data is "too good" and statistical analysis can reveal faked data.

If an error is actually found, the original scientist prints a retraction in the journal where the article appeared, explains the error that took place, and may either withdraw the conclusions or modify them depending on how damaging the blunder is to the original conclusion. If there is no error and the study is actually fraudulent, the repeating scientists will confront the original investigator and ask that the paper be withdrawn. Frauds are often well-meaning rather than nasty people, so they feel ashamed and comply. They are usually not expected to say in their retraction that they out and out faked the data, but instead to state more ambiguously that they are withdrawing the article and that scientists should not cite it. They may also apologize for any inconvenience caused. The purpose of the retraction is not to rub the fraud's nose in the dirt but to get the fraud out of science. Humiliation need not be publicly displayed. Any scientist reading such a retraction understands that the person cannot repeat the work and thus it must be faked. If it is an error, the scientist will identify what the error is so that others will not make the same mistake. Thus, the absence of an explanation involving error implies fraud. If the scientist who committed the fraud stands firm and refuses to admit to the evidence revealing fraud, the repeating investigators will write a note stating that their attempted replication failed and that when they observed the original scientist repeat the work, the work was still not confirmed.

But what if the scientist committing the fraud did not choose an experiment or conclusion that would create a stir? Suppose the fraud wants publications to gain tenure or a promotion rather than fame, or that the faked data supports conclusions already accepted by the scientific community, or that the work is intended for lesser known or foreign journals that few people in the field bother to read? Some frauds have amassed dozens of such obscure publications before they were unmasked by careful readers of the literature. The tip-off might not have been the data itself but an act of plagiarism in which the title was changed, the data was doctored here and there, and the references were altered to look as if it was not plagiarized work. The fraud might have assumed that the original author would never read the obscure journal. It may be possible for such fraudulent work to escape detection forever, and one could also argue that it does not matter because major findings are not contradicted and the work itself is trivial. There are thousands of scientific journals and the vast number of published articles are rarely cited, especially a year or two after their publication.

## The Fate Awaiting Frauds in Science

The two examples I presented actually happened. The first was never unmasked even though the work was important. Fortunately, the work never got past the abstract stage of publication (where the abstract is a short summary submitted when attending a conference), but the physician-scientist did give talks on his work to many universities. He was investigated for fraud at three different universities, but the case lacked sufficient hard evidence because selected patients were used and he stonewalled and threatened to sue when confronted by his accusers. Some institutions do not want to tie up months or years of investigator time repeating faked experiments and spending huge sums of money defending themselves against lawsuits. Prestigious universities do not like bad publicity about their faculty or staff. I knew about the case because someone called me to ask about it. She worked in the laboratory of the accused scientist and she knew that my wife's brother-in-law was a Congressman, the late New York Democrat Ted Weiss, who had taken an interest in medical fraud. Congressman Weiss's staff did not pursue the case because he was less interested in individual cases of fraud than in misconduct by government agencies, pharmaceutical corporations, and other health-providing agencies. He felt that the best protection for the reputation of scientific integrity was careful procedures set up by the institution of the alleged fraud, which would protect the rights of both the accused and the whistle-blower and would be handled without publicity until the case was resolved. When I spoke to him about the case, he deplored the tolerance of such frauds by the scientific community because their claims can lead to injurious or useless treatment that might accelerate the deaths of patients or mislead the emphasis of research in a field, resulting in the delay of more effective treatments. He was particularly angered by the misuse of public funds for fraudulent research, something that university scientists would not tolerate in the private sector if faked data were used, say, to design airplanes or nuclear reactors.

The second fraud, the plagiarist, was a professor who was considered a talented young scientist and hoping to be tenured. His undoing came when an article he had published in a German journal (translated into German by the plagiarist) was recognized by a fellow (rival) graduate student who had studied with the same dissertation sponsor. He recognized the article as familiar and traced it to a review article their professor had written. He then looked at all of his rival's publications and unearthed an article that was published in a Japanese journal (written in English) that had its title and authors removed and the data slightly altered but the written content was word-for-word a published article in a European journal. The rival sent both faked articles (with the originals from which they were copied) to the plagiarist's department chairman. He also sent each editor the appropriate documentation of fraud.

The plagiarist attempted to defend himself by claiming that he had written that section of the review article for his major professor and thus it was legitimately his own words. He claimed that the Japanese article was submitted by a graduate student and that he had just used it as a *droit du seigneur*, or the argument that while it may not be nice, there is nothing wrong with a professor using his student's work without acknowledgement. He refused to name the student because he claimed that she was romantically involved with him. As he was in the midst of a divorce, this would compromise the divorce settlement. When all of this failed and the department chair said that he should resign or face trial from his peers in the university, he threatened suicide. The department chair then called the estranged wife, obtained the contact of the plagiarist's psychiatrist, and got the psychiatrist to convince the plagiarist to resign. The appropriate articles were retracted and no further publicity ensued.

## Applying Today's Standards to the Past May Tarnish Scientists Who had No Fraudulent Intent

Some historians and sociologists of science have looked at the nature of the evidence and data used by famous scientists in the past. Using statistical analysis that was not available a century or more ago, they have found a number of alleged fraudulent papers. Perhaps the first to inspire this endeavor was R. A. Fisher (1890–1962), a respected geneticist and mathematician believed that Mendel's data was too perfect to be true. He charitably claimed that Mendel was too fat to bend over and harvest pea pods, so his assistant must have improved the data for him. The controversy continues and dozens of articles have appeared over the years either defending Mendel or supporting Fisher. I am convinced that Mendel did not conceive Mendelism as a purely theoretical scheme and then faked data to support it. There are several reasons for my belief. We know from Carl Nägeli's correspondence that Mendel sent packets of seeds with predicted ratios to Nägeli, but we do not know if Nägeli planted them. We also know that Mendel received packets of seeds in return from Nägeli and that Mendel did plant them to find that his theory was not universal. Next, today's scientific standards did not exist in the past. In Mendel's day, it might not have been considered improper to count as you go, and as soon as the approximate ratio was nearly 3:1, he stopped counting. A ratio of 70:30 for 100 peas counted does not look as convincing as a ratio of, say, 7507: 2493 for 10,000 peas counted. Much more criticism comes upon certain ratios that should have occurred but did not. When Mendel looked at peas that were either pure or hybrid for yellow, he took samples of ten from a pod and studied those. He did not include the rare event of all ten being pure (the fancy term the geneticist would use is homozygous). Perhaps he did find these, but he may have discarded them thinking that he had a contamination or made some sort of recording error.

The worst scenario about Mendel that I can assume, based on the low likelihood of his inventing the field of Mendelism by some inspired insight into heredity, is that he improved his data by rounding off and excluding some of the data that he did not trust. What is sad about these attacks on the past is that the accused has no opportunity for defense. When a scientist of Mendel's stature is smeared as a fraud, it becomes a headline story and enters the lore of thousands of school teachers who share the scandal with their students. It is akin to what the public does to its national leaders. Presidents since George Washington have been accused of womanizing, corruption, deception, and other moral lapses, making many of them horrifying individuals if the stories are to be believed. At the same time, we know that Presidents, Kings, Prime Ministers, and other high officials are like us and have flaws too. In our zeal to humanize the famous, we are often tempted to believe all rumors about them.

Similar charges have been made about Ptolemy's data for justifying his geocentric universe. Samuel Butler wrote a vicious book attacking Darwin's integrity as a scientist. Louis Pasteur was criticized in his own time and more recently as a less than upfront scientist who did not hesitate to manipulate the data, the truth, or the public. I am not happy with such ad hominem attacks on the character of great scientists. These revelations of fraud are often perceived as unmasking scientists who no longer deserve to be honored. Some go as far as to claim that all scientists are frauds to varying degrees and that there is no such thing as scientific integrity. For this class of sociological investigators, science is no more honest and trustworthy than political position papers, the opinions of newspaper columnists, or a lawyer's brief. By making it a human enterprise in which science constructs its own reality, science is demoted and all knowledge becomes tentative or untrustworthy. We no longer have certainties outside of trivial mathematical tautologies, like $2 = 2$.

## Are Other Scientists' Ideas Fair Game if Not Yet Published?

While outright faking and blatant plagiarism are rare in science, some minor blemishes are fairly common. Almost every scientist can cite a story of someone swiping ideas from casual conversations at a meeting, included in a grant proposal and read by a rival, or prying the information from a colleague or student. I once asked, in a non-belligerent way, a rival geneticist who was a few years older than me (at the time I had just received my PhD) about an article he had published in which a suggestion I made to him at a meeting was not credited at least in a footnote or the acknowledgments section of his paper. He reassured me that it was sheer oversight. He added that it is very common to hear things at meetings and not remember who made a suggestion, and "besides, this all evens out in the long run." He believed that others would do the same and that there was no need to cite or acknowledge a clever idea that came from someone else. I knew that Muller had damaged his reputation because he had a tremendous memory, knew exactly what was said at meetings, and sometimes wrote bitter-toned remarks in his articles that a particular idea of his was swiped. Those who forgot the sources of their ideas or believed that they independently came to the same conclusion without any recollection of Muller telling about it would accuse Muller of having a "priority complex." While I certainly understand Muller's desire for idea priority, I do agree with some of his critics that he often made too much of an issue out of it and indeed earned that reputation. Swiping ideas is much easier to do and get away with than swiping someone's data, faking results, or plagiarizing. Morgan justified his students' unhappiness with Muller's accusations by arguing that ideas are easy to construe and that they constitute such a line of patter by scientists that they should be paid little heed. What truly counts is the well-executed experiment.

# What Gets Left Out in Scientific Papers

A scientific article that reports data from observation or experimentation is not a diary. The scientist must make a judgment call on what should go in or be left out. This judgment involves integrity and should not be a self-serving selection. For instance, every fruit fly geneticist has experienced errors while doing experiments. A cotton plug might be askew and create a passageway allowing flies to crawl in or out. If cardboard lids are used on half pint bottles for cultures of flies in an experiment, a slight crimping of the edge could lead to a happy exit or entry for the flies. Scientists who see such contaminated bottles or vials of flies simply chuck them away. I know of no one who includes in articles the number of dropped and smashed vials, contaminated bottles, or bottles so choked with green mold that it was not worth counting the flies within.

To protect against error, the scientist often repeats the experiment and looks for consistency, and if it is lacking, tries to figure out what went wrong or what change was made that resulted in the inconsistent outcome. When experiments are conducted in a consistent and systematic manner, it is reasonable to believe that the results are valid and will be repeatable by others. Muller often advised us to run "pilot" experiments before doing a full-scale version. He once neglected his own advice and unknowingly made a serious error by having us combine pieces of the Y chromosome of the fruit fly from two different stocks, and the experiment — done on a huge scale — filled the rooms with trays of vials. When the vials were examined a few days later, not a single one contained a fertilized female. All the male flies that were put together from two different pieces of the Y chromosome were sterile. Muller, who was in Europe for meetings at the time, got the message at his hotel and he wired back a telegram: "BIG WORK, BIG QUIRK!"

I am sure every scientist has reflected on the difference between their messy notebooks, cluttered laboratory tables, and failed experiments where the techniques did not work and the finished paper which runs so smooth and logically to its conclusion. In this respect, art and science agree as a process — they go through clouds of error and confusion during formation, but the finished product is what we admire. The difference between the artist's final product and that of the scientist's is the underlying reality.

The artist's is essentially unique. There is no other poem like Yeats' *Leda and the Swan*, but there can be many findings of a 3:1 ratio for two heterozygous parents when they produce many offspring. The work of art is unique to the artist, but the scientific finding is independent of its finder.

For this reason, the fraud cannot claim that science is not objective after all, and that the fraudulent data is just another construction of reality. We expect the underlying reality to be there and the scientific paper to describe some segment of it. Scientists may be empirical, tentative, and open when they investigate a topic, but they do not ignore the reality that they assume to be there and that they hope to find. Scientists are also aware of self-deception, which is why their articles are reviewed by their peers. It is not a perfect system because a rival or someone with a good memory for slights may end up reviewing the article and tear it to shreds. Most of the time the process is fair and the referee is truly neutral even if the submitting author's name is familiar to the referee. A good referee will spot errors in reasoning, shaky evidence, and the need for tighter experiments to clinch a point. The article is frequently sent back with feedback from two or more such referees and the author has to revise the paper, respond point-by-point to the criticisms raised, and satisfy the editor that the responses are to the point and fair.

## Scientific Behavior Can Range from Helpful to Harmful

There are several forms of misconduct that are worth our attention. Some scientists are empire-builders and they like a lot of space and power. They may not be as good a scientist as they claim to be, but they can be successful in convincing those with power that they are that good. They end up forgotten, of course, after they die or retire. At the same time, they are not frauds and they have to publish a lot to acquire that reputation. Sometimes that publication record, which may run to several hundred papers, can be very repetitive, highly co-authored, and of low novelty in the field. Such empire-builders usually do not end up with Nobel Prizes or other high scientific honors, but they end up as chairs of large departments, deans, or other administrative roles in universities or government agencies. Quite a few of these empire-builders are good scientists and they do make innovative contributions and occasionally win significant awards. Professional success and the quality of one's work are not incompatible.

There are scientists who put the names of co-authors on their publications as favors. This practice may be for cognate or unrelated favors, such as the sharing of patient data. Or it could be to help a colleague who is too busy teaching or seeing patients and without a lengthy publication record, the person will not be promoted. While officially frowned upon, this practice has been done extensively over most of the past fifty years. A few embarrassing episodes have led to policy guidelines in journals and universities stating that a co-author must have a significant role in either the writing of the article, the formulation of the basic theory, or the production of the research data. In the recent past, there have been senior scientists of stellar reputations whose generosity in doling out co-authorships was betrayed by fraud. It is tarnishing of a scientist's reputation if a co-author fakes data and the supervising or senior scientist pays no attention and lets it slide.

Who are these frauds? I have mentioned two whom I had personally met with and would never have known of their fraudulent work had not a scandal broke. Other frauds include students (graduate or postdoctoral) who deceive their sponsors with phony data. Since trust is the guiding principle of relations among scientists, there is no obsessive testing of data, but a good sponsor should be able to notice cooked data or errors. A sponsor who is too busy doing other things to know in detail what a student is doing may get blemished by the oversight. Some researchers hope that their assumptions about reality will end up matching their faked data when they finally get around to doing the experiments they should have done. Some defect of personality afflicts such individuals — they are more interested in their ideas than the truth itself. Reality can be very elusive and surprising, often giving us the opposite of what we expect.

Frauds were presented in a "hall of shame" manner during our graduate classes to caution us of what can go wrong. In genetics, the most famous frauds are Paul Kammerer and Franz Moewus. Kammerer was a Viennese zoologist who was alleged to have injected India ink in the thumb pads of midwife toads to claim that the environment had altered the heredity of these animals. Kammerer committed suicide shortly after his work was exposed as fake. Curiously, the novelist and writer Arthur Koestler tried to revive both Kammerer's reputation (he was framed, Koestler claimed) and his theory — the hopeful and recurring belief in the inheritance of acquired characteristics. In his day (just before WWI), Kammerer was widely celebrated for providing the first convincing evidence that the environment directly alters the heredity of animals. Kammerer's life was complex, both politically and psychologically. His personality was seen by his critics as that of a con man. While *The Case of the Midwife Toad* brought out his bizarre romancing of five sisters and his flamboyant personality in Viennese social life, it also tried to justify the legitimacy of the experiments that Kammerer did. Koestler, like Fisher, used the theory of the "other person" to explain the fraud and claimed that it was either a lover who doctored his results because she wanted to help him or a conspiracy of fascists who wanted to frame him because of his communist sympathies. Kammerer had ducked for years when originally challenged and as one opportunity after another to delay was closed off, he felt pursued in the last years of his life.

Franz Moewus was Tracy Sonneborn's idol. When I took Sonneborn's courses on the genetics of protozoa and algae, he spent almost half of one course on Moewus's work. Moewus claimed that he had identified sexuality in a single-celled alga and that a colleague of his had isolated the chemical basis for this sex difference. Moewus' work was suspect in the 1930s, but the rise of the Nazis and onset of WWII kept any independent group from visiting his laboratory or him from coming to England or the United States. Another geneticist-mathematician, J. B. S. Haldane, analyzed Moewus' data and calculated that the odds of them arising from experimentation were remote and that they were most likely fabricated. In the mid 1950s, Moewus finally agreed to come to the US and worked in a laboratory at Columbia University. He was caught fudging his experiments by using iodine to paralyze the flagella of his specimens and he allowed a retraction under his name to be published, dissociating himself from his former work. Moewus then moved to Florida and mysteriously drowned, with some believing that, like Kammerer, he had lost the will to live in disgrace. And just as Koestler had tried to rehabilitate Kammerer, one historian of science attempted to revive the legitimacy of Moewus' experiments in a book that explored the history of cytoplasmic inheritance and the organized hostility of geneticists against it.

Less well known is the story of John William Tower (1872–19??), an American zoologist who claimed to have experimentally demonstrated Lamarckian inheritance in lady bugs. His work was accused of fraud around the same time that Kammerer was accused of fraud in Europe. Tower left the University of Chicago, went to Mexico, and disappeared. Muller said that Morgan and Wilson used to mention in their courses the disgrace of Tower, although I have found no published account of his unmasking. Perhaps Tower, as a less prominent scientist than Kammerer, did not draw publicity on himself and when confronted merely resigned and fled the country to start life anew. Closer to modern times is the case of John Summerlin, a talented investigator and brilliant student in the biomedical sciences who succumbed to the beauty of his own theory that by freezing tissue and transplanting it, he could bring about successful grafts between genetically distinct mammals. His spots of colored fur where the grafts were supposed to have been made turned out to be painted. After his disgrace (the case made the newspapers and gave him a level of notoriety he did not wish to have), Summerlin dropped out of experimental science.

What does this all show? Certainly scientists are subject to the same insecurities as frauds in other fields. Science may be the prototype of the search for truth, but it is also a profession with a positive reputation for its use of knowledge to understand the universe and apply it for good. Therefore, science is respected. However, some people are drawn to science for less than noble reasons. While many end up as competent if not illustrious scientists regardless of their initial motivations, some forget why they are in science and are more insecure — they have a need for glory, promotion, and power, or worry excessively about their reputations or job security. Major frauds are rare as the risks of being unmasked are too high if the claims are startling or exciting. As most individuals with fraudulent intentions soon learn, there are more intriguing ways that nature works than what our imaginations can conceive of. For the lesser frauds, I suspect that quite a few get away with relatively minor sins of self-plagiarism, extending the data of essentially correct work to make it look more convincing, and swiping ideas from others. No one can really determine the frequency of such behaviors because they are not easy to document and, in most cases, it would not be worth the time and energy to build a convincing case. Most scientists, however, are in the field because they love it. They are curious about the world and they want to poke around and explore segments of it. They know that consistent application of legitimate experiments and data gathering more often than not yield interesting insights. The need to fake is a weakness, a behavioral pathology, and most people who have the desire to partake in science are not pathological. They derive pleasure from doing actual science. It would take a very troubled personality to derive pleasure from a life of deception.

# 10 Science and Reality

Few discussions are more likely to make scientists walk away from philosophers than the issue of reality. Philosophers may have debated for millennia what it is other than faith that makes us certain that there is an external reality that exists independently of ourselves. From the scientist's perspective, however, reality is a given. Reality is the material world and the laws that govern it. The scientist is happy to give to whoever wishes it, the world of the paranormal, pseudoscience, the mystical, the spiritual, the hereafter, the soul, or the world of God, gods, saints, angels, witches, ogres, fairies, leprechauns, and the devil. It is not that scientists are saying they don't exist (although some do claim this); the scientist only deals professionally with the real world — the world of matter, atoms, and molecules and how they work on Earth, in living things, and in all observable or potentially observable parts of the universe. If any systems or beings initially assumed to be non-material become manifest by the criteria of science, scientists will be happy to enlarge their universe of reality. Philosophers who remind us not to forget the excluded middle in evaluating choices sometimes forget that there are options other than theism, atheism, and agnosticism. One such option is that of exclusion. If science deals with the material world, then questions about the non-material world are not appropriate for scientific speculation. They only become debatable if the practitioners or believers in these fields or phenomena claim them to be material or real (in the sense of being composed of matter and governed by its laws) or if they insist they are scientific in their practices.

The fields or claims rejected by science include those that are false but claim to be true, such as astrology, alchemy, numerology, phrenology, physiognomy, palmistry, and medical quackery. They also include fields that do not claim to be part of science but have an independent basis for their justification, such as faith and basic beliefs in God, heaven, hell, the devil, angels, the soul, spirits, ghosts, and other non-material entities. A more troublesome area involves fields claiming that certain realities exist, except that we have not detected them yet. The paranormal belongs here, including mind reading (telepathy), foretelling the future (precognition), extrasensory perception, willing matter to move in predicted ways (tele-kinesis), and communicating with the dead. Practitioners in these fields (if they are not frauds or self-deceived) claim that they have had personal experiences or know of such experiences among relatives and friends. They also have specific traditions associated with such paranormal claims. An example is second sight, which is a belief in seeing the present or the future at a distance through a personal trance-like moment when the vision of the unusual event unfolds. The second sight tradition was described by Roman historians visiting Britain.

## Scientists Reject the Idea that the Reality They Depict is Constructed Rather than Described

The category that I wish to focus on is of a different kind, although we will return to it later to examine some of the more traditional ways of describing a non-material universe. There is a growing body of philosophers, histori-ans, and sociologists of science who consider some of the traditional ways that science sees itself as naïve or wrong. Some claim that there is no fixed external reality or at least no way for science to describe it. Others claim that there is some underlying reality, but that reality is not what scientists presently describe. In both cases, there is an implication that science does not describe but rather constructs reality.

Exactly what is this dispute about? As a geneticist, I have no doubts that heredity is largely transmitted by units of heredity called genes, which are organized on chromosomes and transmitted in predictable ratios. Genes have a material reality because they are composed of DNA. The sequence of nucleotides can be used to predict the sequence of amino acids in the proteins that genes specify. Gene mutations can be traced in function from the level of the molecule to the cell, to the tissue, to the organ, to the individual, and to the population.

What I have just described contains factual information and predictions. Thousands of instances of genetic traits can be demonstrated to work this way, and their association with genes, the underlying sequence of nucleotides for those genes, and their mutations have been worked out in exquisite detail. These findings build confidence in a scientist that when genes are discussed, there is a huge mountain of evidence for their existence and how they tie to the chromosome theory of heredity and the theory of the gene. For reasons not clear to many scientists, those who are not actually involved in doing science (such as historians, philosophers, or sociologists) often discuss science as if it were abstract, socially agreed upon, some sort of provisional model, or imposed as a dogma. Opponents of science claim that they are truly describing the material universe, just as I have attempted to do for the world of genes and cells. They argue that what I have described would be no different from astronomers in 1450 who claimed that the universe is moving in epicycles and that the universe moves around the earth, which is plainly apparent to anyone bothering to observe the skies. To some degree, these critics are correct. Before decisive information is available, the data may fit more than one plausible model and there may be no way to decide among contending interpretations, which was certainly true for the gene before the 1940s. Until DNA was proven to be the genetic material, there was no satisfactory way that a gene's shape, biochemical function, or chemical composition could be described. This shortcoming did not prevent a dozen or more geneticists from making guesses about the material composition of genes. Some thought of them as proteins and focused on what they called "autocatalysis" instead of what we call DNA replication today. Some thought of genes as points on a line, having no more reality than a mathematical point and serving a purely abstract function for mapping these hypothetical particles. A few thought of them as beads on a string, some thought the beads were really bean bags, and some thought that the genes were divisible into smaller segments (the Russians called them step-alleles in the 1920s). Rough estimates of gene size were made by studying small deletions or by dividing the chromosome length by the smallest mapped distance between two genes. Not all of these were off-the-wall estimates.

## The Genetics Obtained by Breeding Corresponds to the Genetics by Molecular Analysis

But in contrast to these indirect means of measuring and figuring out attributes, the geneticist's world rapidly changed after DNA was identified as the genetic material. Biochemists entered where geneticists once held a monopoly. When biochemistry proved inadequate to interpret gene structure and function, molecular biology with the tools of physicists entered. When Watson and Crick's molecular model of a double helix solved the problems of replication, mutation, and genetic coding, older methods of doing gene studies melted away. I think that this is true also for astronomy. The Ptolemaic theory becomes silly once the telescope enters and allows us to see the phases of Venus and associate them with the movement of Venus around the sun. The sun's own rotation, seen by Galileo through his mapping of sunspots, makes it possible to think of the sun as a material body. The presence of craters makes the moon as a celestial sphere equally imperfect and material. Most damaging of all are the moons of Jupiter. If the skies are filled with objects revolving around other objects and if the sun is vastly far away and so much bigger than the earth, why does the entire universe have to revolve around the earth? Does not rotation around an axis, like the sun's around its own, explain this more simply? Galileo's reasoning was not based on theoretical principles from which a universe was generated, but from telescopic observations that supported and clarified a Copernican model. What the telescope did to astronomy, the study of DNA did to genetics. Shifting the field from the theoretician's domain to the experimentalist's or empiricist's gives it a material reality that renders much of past speculation false or moot.

## Why Paradigm Shifts are Rare in the Life Sciences

One of the most influential works in the philosophy of science, and perhaps the only such work known to most scientists, was written by Thomas Kuhn. *The Structure of Scientific Revolutions* argued for a process that is easily grasped and on first hearing sounds persuasive. Kuhn claimed that science is usually fairly conservative and abides by a certain agreed upon world view, which he called a paradigm. In 1450, that paradigm was the Ptolemaic view of the universe where the earth was the center of it all. Both common sense and religious tradition made this a reasonable model. Copernicus, of course, challenged and changed that paradigm with one of his own. New evidence is not needed to bring about the scientific revolution called a paradigm shift. It is often a new way of describing the same thing by shifting things from old to new categories and sometimes renaming them. Thus, for the Ptolemaic universe, the earth was not a planet; it was the center of the universe. The sun, moon, Mercury, Venus, Mars, Jupiter, and Saturn were all planets. The stars were not planets because they were fixed in space with respect to themselves, but planets were bodies that moved to different parts of the sky. We might think of the Ptolemaic system as the earth-planet-star model. Copernicus made the shift by moving the sun to the center of the universe. As a result, the earth became a planet and the moon became a satellite of the earth. We can then think of the solar system as a sun-planet-satellite-star model. Making the sun a star required an additional shift that would make it consistent with Galileo's telescopic studies, but before his observation of sunspots and the reflected light of the planets, it would have required an assumption by Copernicans that borders on heresy because switching places for the earth and the sun would demote the earth and raise the sun to the center of the universe.

Scientists would not find fault with this part of Kuhn's analysis. There are theoreticians who rearrange the way we conceive things and in rewording them give us new insights. This was certainly true for Schrödinger when he renamed the gene's specificity as "the code script" and the gene's ability to reproduce its variations as the "replication of an aperiodic crystal." By shifting from the descriptive language of biology to the descriptive language of cryptography or physics, Schrödinger created a paradigm shift.

What is more difficult to accept is Kuhn's thesis that within a paradigm, normal science, as he calls it, prevails. Kuhn sees normal science as puzzle solving. According to him, scientists do not challenge the paradigm; they accept it and work to fill in the details. This view reduces scientists to less able, less imaginative, and somewhat stodgy people. The real winners in Kuhn's world of science are the paradigm shifters, those with a bold vision for reorganizing knowledge. Experimentation and discovery are reduced to a lesser role, one of annoyance when the paradigm is pushed and cannot accommodate new findings and that of the technician's work — filling in the details of the broad picture that the theoreticians, as paradigm shifters, provide.

It is difficult to find many examples of paradigm shifts in the life sciences. There are five great concepts that collectively interpret the life sciences. The first is the idea of cells as the basic units of life. The second is the idea of the gene as the unit of transmitted inheritance. The third is the idea of a life cycle, that we change from conception to death. The fourth is the idea of molecular biology, that there is nothing other than physics and chemistry that determine the material basis of life. The fifth is the idea of evolution, that all life undergoes constant change and that species arise gradually, change, or become extinct.

If we examine all of these ideas, we find it hard to identify a single paradigm-shifter. As we saw, the cell evolved as a concept from empty boxes to cells filled with a nucleus and cytoplasm. No competing model was replaced. The gene was an idea that evolved out of hereditary units, which was itself an idea that began to emerge in the mid-nineteenth century, inspired perhaps by other unitary models such as atomism and cell theory. There was no contending physical model of heredity. The life cycle was observed and applied to smaller and smaller organisms as the techniques to do so permitted. Evolution did not replace a scientific theory; rather, it replaced a religious view of special creation by a deity whose closest legitimacy as a science would have been natural theology. Natural theology was virtually gone from scientific thinking before Darwin's *The Origin of Species* appeared. The secular language and needs of the industrial revolution and its supporting sciences likely made references to the Creator less important in understanding the basic sciences that industry relied on.

## There are Differences in the Way Social Sciences and Physical Sciences Approach Objectivity

Kuhn never claimed that there was no underlying reality described by scientists, but some of his enthusiastic supporters in the social sciences found that idea appealing. Both history and sociology are fields that recognize the difficulty of objective reality in their disciplines. History is always incomplete and sociology describes societies which are, by their nature, transient and constantly changing. Thus, trying to locate reality in a kaleidoscopic world becomes difficult if it is possible at all. Both history and sociology establish their theories on populations of people and the groups and values they hold. Individuals vary widely in personality and one cannot guarantee that the personalities of one generation will be similar to a future one. This variation is less of a problem in science where individuality may be relatively absent at the level of the atom or molecule and where statistical measurement is done with behaviors at the level of Avogadro's number (that barely conceivable figure of ten to the 23$^{rd}$ power!). In physics, all falling bodies in a vacuum, with trivial exceptions, would show the same acceleration as they fall. The life sciences, however, do show some kinship with history and sociology because there are millions of species and virtually no two individuals are genetically alike within any one species. One could also argue that the genes of today's generations are unlikely to have the same frequency and distribution across the continents several generations from now. Why then should biology not be considered in the same boat as sociology and reject the presence of some underlying reality? As I tried to show, when working with science, even for the mutating gene, there are certain constants such as the retention of the copying property by mutated genes, the retention of DNA as the basis of making a virtually infinite combination of sequences over time, and a much longer persistence of basic life processes (all five of the basic concepts apply to life some 500 million years ago as they do to life today).

## Scientists Feel More Comfortable with Popper's Idea Of Falsification

The second contender for shaping the philosophy of science in the twentieth century is Karl Popper. In numerous articles over a long career, Popper argued that a scientific principle is always tentative and cannot be proved. Science earns its legitimacy, he claims, by allowing its views to be falsified by contradictory evidence. If a theory lacks the capacity to be tested with a possible contradictory outcome, then science is no different from dogma or faith. Certainty and truth, in Popper's view, do not belong to science.

Many scientists like this idea and see nothing wrong with it, even though they have never been taught such a credo as young scientists.

There are difficulties with this belief. It verges on nihilism, which is a belief in nothing, and it allows all contending models to claim legitimacy until they are falsified. Some fields, like the paranormal, posit a non-material or undiscovered law that governs reading the future, mind reading, or some other similar phenomena. Advocates for the paranormal might claim that they are open to discarding their belief in extrasensory perception if someone can falsify it with adequate tests. In this case, adequate tests might mean not using skeptics whose hostile thoughts might interfere with the ability of the psychic to receive transferred thoughts from a distance.

Popper's theory also makes it difficult for fields like evolution (biological and geological) to claim legitimacy. How does one falsify a claim that natural selection over tens of thousands of generations may lead to a divergence of species? It is both the glory and the disappointment of philosophy that Popper's reliance on the power of consistency, definition, and logic makes it easy to expose poor arguments and at the same time support arguments that trivialize or reject science. I find it hard to think of the cell as a tentative theory subject to falsification. It is equally difficult to think that something other than DNA is acting as the genetic material for the thousands of traits that have been described, whose genes have been isolated, sequenced, and cloned, and whose predicted products can be produced when the genes for them are inserted into viruses, bacteria, or synthetic vectors.

## Scientists Prefer Objectivity to Truth When Describing Their Findings

Exactly what is meant by proof or the term "true" when one discusses science? Science is not a tautological equation. The counterpart of a gene or a cell is not a complex set of mathematical equations. To reject the truth of well-confirmed experimental evidence in science suggests that it is also legitimate to reject anything we hold to be true about the material universe and that only platonic ideals, non-material beliefs in souls and other spiritual phenomena, or abstract mathematical and logical concepts would be true. If truth is stripped from all of reality, then it is difficult to avoid the impotence of nihilism, unrequited skepticism, or a lapsing into solipsism, the ultimate belief that only the self exists and the entire universe is one's own projection.

It is for this reason that scientists have hostility to philosophers who tell scientists what they are really doing or what constitutes legitimacy and distinguishes science from superstition, confusion, and deception. I say this not to exempt science from the toughness that sloppy thinking deserves. I argue instead that science approaches the problem of objectivity and truth in a very different manner from traditional philosophic theory. The scientist strives to attain not so much logical consistency but rather consistency of evidence. For a cell theory to be valid, samplings of all taxonomic categories must be done. The scientist relies on statistical evidence of reliability and for most experiments and findings, a reliability of 99 out of 100 is excellent and 95 out of 100 attempts is considered acceptable. Most scientists look for both that statistical outcome and the more convincing finding of repeatability when an experiment is done again, especially by someone else or when the basic finding is extended to other organisms or situations. Let us assume that there are about 2 million species of animals and that there are about a dozen major phyla running from coelenterates (jellyfish) to vertebrates. There may be a dozen or so major classes within each phylum and each class may in turn have a dozen orders. This pattern gives some idea of the splitting into major groups. How would one approach this issue of the cell theory? Testing every species would be an almost hopeless task because not all species have been identified, but scientists can make a broad assumption based on consistency. If cell theory is universal, why not choose a few classes within each phylum and a few orders within each chosen class and work it down so that some five or ten species in each of these orders is tested for the presence of cells in its tissues? For the class of mammals within the phylum of the vertebrates, there are lots of orders. We belong to the primates; we could choose a carnivore like a dog or cat, an ungulate like a cow, an edentate like an ant eater, or a rodent like a mouse. Then if all species within a variety of orders among a variety of classes within a variety of phyla all show that they are composed of cells, our theory has been vindicated. The scientist does not have to show that every species is made of cells. One does not have to match Avogadro's number to convince oneself that something is uniform or nearly so.

## Diversity in the Life Sciences Makes Universal Claims Unlikely

If we did this study and included phyla that were not Metazoa (multicelled animals), we might run into some unusual findings. We might find some organisms that are not composed of cells but are large microscopic entities with more than one nucleus, many complex organelles not found in metazoans, and sophisticated behaviors with complex life cycles (such as protozoan parasites which have many stages in different hosts). Does this mean that the cell theory is wrong? No; no more than the finding that bacteria lack the organelles usually found in metazoan cells. Nor does the absence of cellularity of any kind in viruses make them less of an organism. What we recognize, especially in the life sciences, is the diversity of life and the way life was built on itself and sometimes strips itself down or compromises an essential principle for its own survival. Thus, in a human, there are times when cells do not exist in a typical way, such as a nucleus bounded by a contained cytoplasm. Instead, we might get a pudding of cytoplasm without subdivisions whereby lots of nuclei are present, which can be seen in voluntary muscle cells or the implanting layer of the blastocyst when it digs into the uterus. This is called a syncytium. In biology, exceptions do not necessarily negate a theory; they may reflect adaptations. All human syncytial tissue is derived from perfectly normal cells that are traceable to the fertilized egg, which is also a cell, albeit a big one. We think of a hand as having five fingers, but we do not deny that a baby has a hand even if a sixth finger is present. Definitions may trouble philosophers who seek consistency, but they do not trouble a biologist who is aware that there are exceptions or modifications to almost every defined anatomical structure.

In a way, precise definitions are platonic ideals. In life, there is a cloud of ambiguous extensions and contradictions to every definition. But rather than rendering the definition useless or wrong, the presence of ambiguity means that precise definitions exist only in a platonic world, not a material world where complexity and diversity rules. Nevertheless, definitions are good enough for a scientist to recognize a hand regardless if the baby is born with syndactyly (the fingers are in a mitten-like common wad of flesh), polydactyly (there are more than five fingers), ectrodactyly (the hand is split down to the wrist making it look superficially like a lobster claw), or brachydactyly (the fingers and thumb are squat and thick).

# 11 Science and Industry

Most scientists who receive the PhD as their certification of being a competent research scientist seek an academic position in a university where they will teach and do research. They are hopeful that they will get to do basic research on whatever interests them and learn something new about the universe. They tend not to be interested in applied research where they would be working on some commercial or military applications of basic science. A much smaller number who desire to do research full time but not teach will seek positions at a research institute. There are far fewer of these than there are universities with PhD granting departments. Even fewer are the PhDs who end up doing basic research in the think tanks of major corporations.

Most young scholars are often surprised to learn that around 90% of those who receive a PhD will never publish anything other than their dissertation research. There are over 3,000 colleges and universities in the United States and less than 200 are major universities that grant PhDs and have active research programs. Those who end up in colleges that do not grant PhD degrees or relatively modest universities will not have the opportunity to do research. Instead, they will be expected to spend most of their time teaching undergraduates. Their research activities will be limited to undergraduates and most of that work will not be publishable because it is a part-time commitment and they are cut off from the day-to-day stimulation of working with graduate students, postdoctoral fellows, and colleagues actively engaged in research. They will not be bringing in federal or even private grants to subsidize their research, and most colleges have inadequate budgets for first-rate research.

## University Science Leads to Tenure and a Long Commitment to Research

Life in the university or college is relatively secure after tenure. Tenure is granted to an accomplished scholar who advances in rank from assistant professor to associate professor. With tenure comes academic freedom and job security. The associate professor (and later on, when even more distinguished, the full professor or just plainly, professor) will now call the shots, decide what research is interesting, and can take risks by exploring new or controversial topics. Academic freedom, in principle, allows the tenured professor to criticize policies of the government, the university, or society without fear of being fired for unpopular or controversial views. This is a major reason why scientists who are interested in pursuing some aspect of reality (whatever philosophers or society may consider that to be) prefer an academic tenure-track position. In almost all other forms of employment, academic freedom does not exist, which is why employees in those companies often look the other way when abuses or questionable practices are observed. Being a whistle blower or a critic of one's bosses is a quick route to being fired. If a scientist is muzzled by such fears, then objectivity and integrity might be compromised.

Despite this preference for academic freedom, there are students who want to work in industry because they like applying what they know, they want the higher salaries that industry provides, they cannot stand teaching students, or they know that they would have a hard time finding a tenure track job because they did not receive their PhDs from a first-rate investigator or institution. Clearly the PhDs of a Nobel laureate will have first dibs on available jobs and are most likely to end up in one of the desired universities with a tradition of supporting faculty research. During the Cold War, there were many jobs available in defense industries, but there are fewer of those today given that the cycle of human affairs has largely shifted to a peacetime economy. It is sobering to think that the twentieth century had 1901–1914, 1919–1939, and 1989–2000 as peacetime years whereas the remaining 55 years were either engaged in war or its preparation. Even with the end of the Cold War, which began almost as soon as World War II ended, we are by no means in a peacetime economy like the one between the two world wars or prior to World War I. The momentum of government subsidy for defense industries is enormous because too many people would be unemployed, flooding the market with skilled job seekers. We still maintain huge defense commitments even though the justification for them is less convincing than when we had a Cold War.

## Secrecy is Incompatible With Basic Science

Both industrial research and military research involve some level of secrecy. A commercial process or invention may require a considerable investment of money. The work that gives a company competitive advantage has to be protected, and it may take several years to secure patents. The corporation may also want to gear up and get its product ready for marketing before it announces its novelty. Competitors will quickly jump in on something new, desired, and profitable. Thus, employees are required to sign contracts that demand their silence and they are not allowed to take ideas in progress with them if they are hired elsewhere. Their contract makes them and the company that will employ them legally liable to expensive lawsuits if they divulge or use trade secrets.

Secrecy is shunned by academic scientists because a free flow of ideas is more beneficial than an atmosphere of seclusion and silent competition. At any scientific meeting, there is an enormously invigorating exchange not only during the give and take of the question and answer period following a presentation, but also during the coffee breaks, dinners, and lounge conversations. Scientists like to share about what they or other laboratories are doing. Such exchanges are a means of showing respect for one another and appreciating a common value of science — the desire to know. As in all endeavors, there will be those who are secretive, fearful, or manipulative, but the possible presence of such toxic individuals does not and should not discourage scientists from engaging in synergistic openness. Almost all scientists come back from meetings feeling exhilarated and eager to try new approaches to their problems and use new techniques learned through these informal channels.

This openness is necessarily absent in industry. Scientists in one corporation cannot eagerly divulge their ongoing projects and benefit from the stimulus of a free exchange of ideas. They need permission from a project director and that permission will be tied to corporate interests.

But this practice is not necessarily wrong; it is just a different way of doing science based on a different set of values. However, this is also why few Nobel prizes have gone to scientists in industry. The overwhelming number has gone to those who work in universities or non-profit research institutes like the Fermi lab, Cold Spring Harbor, or the National Institutes of Health. Only those industrial think tanks that grant something equivalent to the academic freedom of the university have enjoyed that stature for its scientists, such as the Bell laboratories.

Perhaps the most difficult value associated with industrial or military science is secrecy. Muller told us that he declined participation in any project that required his secrecy and he declined receiving any information that was to be given to him in confidence. He said that it was impossible for him because knowing something true about the universe and yet marking it as secret and not using it paralyzes thinking. You cannot go on as before with an erroneous view. You cannot develop new experiments and approaches to scientific work that needs that secret knowledge and thus it becomes frustrating. I followed his advice when I came across a packet of letters that Muller had written from Stockholm to his wife at the time of his Nobel Prize. She was too ill to accompany him. I turned over the packet of letters to Mrs. Muller without reading them. She came from a tradition that believed that letters to her from her husband were a private matter. I realized that I could neither quote from such letters nor paraphrase the information gleaned from them without revealing my source, so I had to choose between betraying a confidence and value that she held important or putting into my biography of Muller an eyewitness detail and expression of emotion that could not be found from any other source. I chose to respect the values of the person who loved her husband and who valued the privacy of their intimate comments.

There is no question that secrecy delays or deforms science. When Charlotte Auerbach discovered the mutagenic effects of nitrogen mustard and mustard gas in 1940, she had to wait until 1946 to publish. It is not clear why her work should have been labeled secret. Mustard gas was used in World War I and it produced horrible burns and was known to be radio-mimetic, which was why the pharmacologist J. M. Robson suggested to Auerbach that she should study the compound (he was a co-author of her first paper on chemical mutagenesis because of his suggestion). But in war time, the role of the censor is to put everything that has to do with military usage, past and present, under a lid of secrecy. Had mustard gas never been used in gas warfare, Auerbach would have published five years earlier.

## Many Scientists are Motivated to do Science for the Public Good

Many scientists see the applications of science as contributing to the public good. There are new medications and contributions to longevity and public health for those in the life sciences. There are lots of agricultural applications of genetics, physiology, and developmental biology that produce billions of dollars in profits, employment for hundreds of thousands of people, and an abundance of food and varieties of food all year round. Chemists, physicists, engineers, and geologists can all claim similar contributions to human benefit. These contributions underlie part of the positive public image of scientists.

Why then would scientists lend their talents to military applications? Weapons kill not only warriors but civilians and most modern weapons are so destructive that it is hard to prevent the killing of innocent people. Artillery shells lobbed at cities ten or more miles away destroy not only tanks but buildings that happen to be nearby. Squadrons of bombers that drop thousands of bombs on a city know that a considerable number will miss the railroad tracks and bridges or defense factories and blow apart churches, orphanages, hospitals, schools, and the homes of tens of thousands of non-combatant civilians including children. While religious values teach "thou shalt not kill", there is a subtle barrier between church and state that churches rarely cross. They cannot voice criticism of the government that tolerates or promotes their freedom without risking accusations of subversion and oppression. A few pacifistic churches are tolerated because they preach non-violence and make it a religious mandate not to kill. Most churches rely on individual conscience, not scriptural absolutism to guide their faithful. Too often, churches of almost all faiths in all countries look away from the evil done on a massive scale by those who work in defense industries. This is also true of those who work for these industries in whatever country they happen to live in, choosing patriotism and defense from enemies as values. Their work is seen as necessary to keep the country from invaders and to preserve cherished peacetime or national values. The combination of patriotic values, a good salary, enjoyable work conditions, and focused projects takes away many of the insecurities faced by basic scientists who do not know if their projects will work or not, who do not get as good a salary, and whose financial support for doing science has to come from federal or private grants. But there are also disadvantages for scientists who work in the military besides an uncomfortable conscience. Governments are fickle — they appreciate and foster patriotism, but they can quickly withhold their subsidies from the industries that provide them weapons when it is politically expedient to do so. Over the years that I have taught, I have seen the impact of these political changes, with engineers who once worked on missiles and supersonic aircraft for local military industries returning to school to become high school teachers or finding some other employment with a new degree.

No one has solved this universal problem. People will continue to believe that they are doing good by making weapons to kill the innocent. They will deny the reality that they are, have been, or will be in some way connected with the deaths of innocent people. In all countries, a common response is often heard: We live in an imperfect world, not a utopia. Until we are in a universal state of peace (such as that imagined by many evangelical Protestants after the second coming of Christ), there will be wars and it is our duty to help our country in war or before it is engulfed in war to defend itself against its enemies.

## Support for Science Influences the Kind of Science That can be Done

In the first half of the twentieth century, colleges and universities had little connection to industry and government. Research was largely funded out of the investigator's pocket or supported by the university from its endowment funds or from the very few foundations that took an interest in basic science, such as the Carnegie Endowment or the Rockefeller Foundation. Scientists looked somewhat suspiciously at graduate students who received PhDs and went into industry, and they looked equally suspiciously at colleagues who moonlighted in the evenings or during summers by using their skills for industry. Some schools had strict contracts disallowing such consulting arrangements. Some did allow them, but the earnings were either limited or a portion of it had to go to the university, especially if it involved patents on products or processes that emerged from research done at the university.

After the launch of the space satellite Sputnik, the federal government panicked and realized that science does not flourish only in a democracy as it had deceived itself into believing. The Soviet Union not only had nuclear weapons but also sufficiently powerful rockets to launch a small spherical, beeping satellite that sent its signal to radios and television sets across the US as it passed by. Engineers immediately saw that such rockets would soon be packed with nuclear weapons and that it was now possible for a country thousands of miles away to lob nuclear weapons from its own territory. Within a few years, the university changed. The National Science Foundation and the National Institutes of Health expanded their support for graduate students, and I was well supported after my first year. The NSF Fellowship that I had in 1954, which was hard to get, was replaced by the more easily attained NIH Fellowships which I had until I graduated with a PhD in 1958. Muller no longer depended on Rockefeller Foundation grants for his busy laboratory; he was now easily the recipient of NSF and NIH grant support. School children were soon reading the yellow, green, or much-coveted blue versions of high school science texts that differentiated students according to ability levels. Science was considered "the endless frontier" and huge sums of money went into subsidizing both basic and applied science.

Some universities recognized an opportunity to use university talent to attract and establish a ring of industries around them. MIT and Stanford were the first leaders of this movement. Money from industry could now provide graduate students with summer stipends, job opportunities, and a chance to work with expensive equipment and learn how to apply their basic science skills. Some claimed that their research was really basic science and it was simply coincidental that some of that knowledge spilled over into industrial applications. In the 1960s when I was still at UCLA, we used to debate these intrusions of industry — both private and military — when they were offering support for students in return for the opportunity to hire them after graduation. I cannot think of a single biochemist then that looked favorably on these overtures from industry. I remember Nobel chemist Willard Libby asking us at one of our departmental meetings to participate in such an industrial venture with the university. When he received a polite silence after his pitch, he got angry and said, "You're damn fools if you don't take this opportunity. You can be sure someone else will."

Today's academic world is different. Both private and public universities are eager to set up partnerships with industry. State legislatures and local industries like the arrangement, students fancy the employment possibilities, and faculty enjoy having grant support and the opportunities to consult and make money from their contributions. Despite the academic freedom in the university, there is very little opposition to these overtures. Neither does anyone, including myself, want to be self-righteous and tell colleagues that they are selling out for money, nor does anyone want to irritate a state legislature by saying, "No thanks, we don't want this applied venture with industry. Our mission is basic science. But please support us anyway." As a result, there is increasingly a look the other way attitude. If students are testing a product for a pharmaceutical company, everyone pretends that this is basic science, or at least a process of learning tools that can be applied to basic science. Students want summer jobs and they are not going to discriminate too carefully between applied and basic research. Tenured faculty may decide to focus on industrial rewards and applications and do not feel that they are compromising either talented undergraduate students or graduate students by shifting their basic science research toward something that looks commercially interesting. Many today will be sympathetic to Libby's comments.

## Scientists Must Balance Their Values and Their Opportunities to do Science

I am neither a pessimist nor a drudge. I do not believe science has sold out to industry and has lost its capacity to do basic science. Scientists are very curious about nature and many will not leave their lower paying university positions or sacrifice their time for research for consulting work or summer income from industry. They will continue to enjoy doing what scientists have done for centuries in democracies, under tyrants, or even under powerful opposition, as Galileo himself demonstrated in his own time. I mention this to make both scientists and the broader public aware of the values they have to contend with while shaping their careers and the powerful influence of society in providing support for science. No basic scientist wants to do away with applied science. The most talented scientists will still choose to work at research universities and seek the satisfaction of academic freedom rather than the restrictions that must necessarily be imposed by applied science.

# 12 Science and Personality

There are popular images of the scientific personality. One is negative and portrayed in countless movies, comic books, science fiction stories, and journalistic commentaries — the scientist who is so carried away by the intoxicating power of knowledge that it leads to disaster. It is as old as the story of King Midas's touch, recurs in Marlowe's *Dr. Faustus*, and manifests with less evil but nevertheless still carries an unsettling personality in Goethe's *Faust*. Mary Wollstonecraft Shelley's *Frankenstein* is the quintessential prototype of the scientist who plays God. The contrasting public image of the scientist is the selfless seeker of the truth. My favorite candidate is the titular character of Sinclair Lewis's *Arrowsmith*. This type of scientist is the humanitarian — the dedicated, modest, and generous provider of new and useful technologies, like Pasteur giving away all his discoveries without patents for the good of the public.

I doubt that I have ever encountered these two extremes of the scientist as conceived by the public, but I find that the real life scientists I have met or whose biographies I have read paint a more complete and interesting portrait. My experience is necessarily heavily tilted toward those in my own field. My readings of biographies of noted physicists, astronomers, or chemists tell me that the personalities I describe are not unusual and that a similar spectrum exists in all fields of science. What they illustrate to me is the independence of discovery and invention from individual personality. I begin with Thomas Hunt Morgan, Muller's teacher and the founder of what has come to be known as the fly lab, the Drosophila group, the Columbia school, or Morgan's gang, depending on who is describing it and their relation to the group. Morgan was the son of Charlton Hunt Morgan, one of two Confederate Generals (Morgan's Raiders) who led commando attacks in Kentucky and Indiana. They were looked upon as bandits, horse thieves, or terrorists in the North and as liberators and heroes in the South. Morgan's father had been active in politics before the Civil War but was persona non grata after the war and had to settle for a life growing hemp for the sailing industry. Morgan's family must have taken a lot of patriotic pride in his ancestry, which descended from the Mayflower (J. Pierpont Morgan the financier was a distant cousin) and included Francis Scott Key as a great grandparent on his mother's side. T. H. Morgan hated history, genealogy, and patriotic puffery. He declined going to dedications to his father and uncle throughout the south and remembered with loathing the parade of "living in the past" veterans who descended on his father's hospitality.

Morgan was one of the early Johns Hopkins PhDs who found in the newly imported European tradition of scholarship and experimentation his life's commitment. Morgan saw in experimental biology the only legitimate way to move biology away from the tradition of mounting museum exhibits and dwelling on classification. When he achieved fame for the flood of papers he wrote on experimental embryology, he was recruited to Columbia by Wilson, a somewhat older graduate of the Hopkins school. Morgan called himself Professor of Experimental Zoology. He despised philosophic approaches to science and was a committed materialist. He was skeptical of all theories until they were experimentally tested. Muller called him a doubting Thomas and "doubted the doubt until he doubted it out." Morgan preferred a casual approach to life and dressed in a sloppy manner, sometimes substituting a piece of laundry rope for a belt to hold up his pants and going off to meetings with his shaving gear and toothbrush wrapped in a paper bag and stuffed in his overcoat pocket. He played the violin and enjoyed fiddling, sometimes playing with folk singers when he made trips back to Kentucky. He was witty, enjoyed a free exchange of ideas with his students, and treated them like colleagues or an extended family. He also played favorites and adored Sturtevant above his other students, much to the seething sibling rivalry of Muller who felt like the younger brother who was left out of the bounty that was showered on the first born. Altenburg, Muller's ally in this not quite happy family of classical geneticists, entitled his obituary of Morgan as "T. H. Morgan, Democrat." At a time when relations of faculty to students were often formal, this ease of respecting all as co-workers and inviting them to enjoy science as a pleasurable experience was formative, even on those who felt resentful over their perceived inequalities. When there were celebrations like publishing an important paper or a graduate student becoming engaged, Morgan would prepare a Neapolitan squid dinner sautéed in olive oil and washed down with robust Italian wine. He asked his students to come to his apartment once a week, where they would take turns reading from current books and papers of interest. He protected Calvin Bridges, an orphan and a free spirit who believed in free love and who later declared himself an atheist in Who's Who. He tolerated Muller's difficult personality and took him on as his student despite Muller's many outspoken criticisms of Morgan's working style and occasional inconsistent thinking. Morgan had a stubborn resistance to Darwinian natural selection and sought futilely for a model of evolution that would bring species into being by some other means, especially through the type of sudden "jumps" that Hugo De Vries favored. It was that search for new species by experimental means, which De Vries erroneously thought he had found, that made Morgan switch from embryology to genetics. His attempts at breeding fruit flies was stimulated by the hope that he would luck out as did De Vries and see some new species if he bred the flies in large quantities and long enough, a task to which he devoted almost two years before stumbling into the less dramatic mutations that gave him X-linked inheritance, crossing over, and the theory of the gene. Morgan dumped ideas and theories with no regrets; he went with the experiment and had little regard for grandiose speculation.

Although Morgan is justly famous as the founder of classical genetics with the new principles it gave us, he is underrated for a major contribution that he pioneered. Morgan was the first major scientist to develop what is now taken for granted — team research. Throughout the nineteenth century, research was done by individuals working alone and published almost exclusively in their own name. Morgan's papers were frequently identified with his students as co-authors, a tradition that was virtually non-existent in European universities and rare in the US. By making his students colleagues in the pursuit of common goals, Morgan inspired hundreds of scientists in the life sciences and in other disciplines to follow his lead. This practice became the hallmark of American success in graduate education and reversed the trend that Morgan himself had followed. By the 1910s,

Europeans were coming to Morgan's laboratory to study genetics. American science was developing a worldwide reputation and experimental science became firmly rooted in every major university.

## How Morgan's Student, Muller, Differed in his Scientific Career

Muller was much more reserved and appreciated the role of theory and speculation leading to experimental testing. Ideas intrigued Muller and he generated many of them — sometimes from crumbs of data — and built them into spectacular models that often met the test of experimentation. Muller grew up in New York City as a third generation American of German Catholics, who abandoned their religion and favored socialism when they fled from the collapse of the revolution of 1848. His father married a part-Jewish Episcopalian woman, so Muller was of mixed heritage, atheistic, and occasionally Unitarian in his upbringing. He became, like Bridges, a committed atheist by the time he was a college student and he was the most zealous of the socialists in Morgan's laboratory, switching later to the bolshevists after the Russian Revolution. Muller knew poverty after his father died while Muller was still in elementary school and the meager income from a partnership in a metal art shop was their sole means of support. He was envious that other graduate students could depend on family support or get stipends to support their work while he had to work part-time, rush from class to job and back, and do as much laboratory work as he could muster without fully destroying his health. Muller also chose the path of the theoretician in those days to compensate for his lack of free time to do experimental work. He rankled with anger when he saw his ideas frequently end up as the basis for experiments carried out by others and his name would not appear as a co-author. From his graduate days on, he would be obsessed with priorities and would not hesitate to remind others of the date when he first conceived of an idea credited to others. Muller often mistook criticism of his work with criticism of himself personally, which is a damaging personality defect in all occupations but especially painful for a scientist who must constantly face the criticism of peers.

Muller was intense, totally committed to the truth of science, and told us that we should not be in science if we thought it was just a game. He demanded a seven-day work schedule and got it from all of us who worked as graduate students. How could we not when he was there late in the evening on December 24, writing reports and providing critiques of papers for colleagues elsewhere who imposed on his time? Even when he was committed to his bolshevist social views in the 1920s and early 1930s, he felt that the philosophic treatment of science by socialists was weak and not worthy of respect. Muller's brilliance was acknowledged by all who heard him give lectures at meetings or who engaged in conversation with him about their work. I never met any other scholar who worried over every sentence his students wrote and made sure that no sloppy thought or unexamined idea slipped by in their writing. As a lecturer, he would often drift into tangents due to the cascade of thinking on his feet, inventing as he went along and reliving the past while trying to convey the history of an idea. Muller's style was to the displeasure of some of his students who enjoyed the more polished and organized lectures of other faculty, but I felt that every single one of his lectures was a banquet of information, ideas, history, anecdotes, and insights that never failed to give me goosebumps.

Muller kept his personal life private and did not entertain much at his home. He had a sympathy for the underdog and a secret generosity that was at odds with the poverty-driven values that made him save odd shapes of paper for writing, weigh rubber bands to get the best buy for his laboratory, and serve cheese dips when he did entertain. I learned years later that when his fruit fly stock keeper, an African American woman who was supporting her husband through graduate school, had her first child, he gave her a check for $500. Muller also paid for a new suit for C. P. Oliver, one of his students at Texas, as he believed that appearance is important for a job interview and Oliver could not afford one during the Great Depression. When I wrote Muller's biography, I came across other instances where Muller's checks went to needy scientists and students whose plight he absorbed and tried to remedy.

I was impressed by Muller's conscientious efforts to answer all his correspondents' requests. Only rarely, with what he called "nut mail," did he sometimes not respond. But critics of his views on eugenics or radiation policy, whether they are famous or virtually unknown, received his honest replies. If supporters of a cause tried to lay a guilt trip on him for not joining their cause, he would let them know that it was his life and his choice to live it as he saw fit. He rejected any compromise on public safety when he served on commissions and he refused to let Cold War politics sacrifice public safety out of political expediency. He was condemned both by the Soviet Union as a spy sent to subvert their science as well as by some of his critics in the United States as a Soviet spy sent back to subvert our nuclear weapons program by expressing concern over low doses of radiation. Muller stuck his neck out on issues where he had taken a firm position. He told us that we would find ourselves facing similar controversies in our careers and that it was the nature of genetics, which is so close to what matters in people's lives, that makes almost all its findings subjects of controversy.

## Max Delbrück Preferred a Guru's Role in Science

If Morgan was relaxed working in genetics and Muller was intense, Max Delbrück had the reputation of a guru. He was quite happy to pull out rather early from the field he founded, bacteriophage genetics, and let his students and colleagues divvy up the problems and go after them. Delbrück saw his role as a goad. He preferred original and difficult work, not the ongoing solving of details. Onwards from the 1950s, Delbrück committed himself to cracking the relationship between photons of light and plants (especially microscopic molds), which responded with movement. Exactly what made non-photosynthetic water molds receive photons and process them such that they act as signals to set off plant or cell behavior was doubly difficult to interpret. Unlike green plants, which have chlorophyll as a natural receptor to light, water molds lacked such a system. Delbrück spent some 30 years trying, without success, to solve this problem. It was his way of paying homage to Neils Bohr, whose lecture on "light and life" got him out of atomic physics and into biology.

Delbrück was notorious for his apparent rudeness. Many a speaker mentioned giving a talk at Caltech and dreading the moment when Delbrück would get up and say, "Without doubt this is the worst seminar I have ever heard." I consider myself lucky. When I gave a seminar on my work involving mosaicism induced by chemicals, Delbrück was polite and did not insult me. I once greeted him at the Athenaeum Club at Caltech and he asked, "Muller is always sending me reprints but it's all this eugenics and radiation hazards stuff. Why doesn't he do science anymore?" Another time when I was in an audience at a lecture he gave at Caltech, he mentioned approaching the problem of gene size by using data on crossover frequencies. I cautioned him that this could be difficult because crossover frequency varies depending on the segment of the chromosome. A few weeks later when I was back at Caltech, he pointed to me and said, "I was told that your objection isn't valid; were you trying to pull a fast one on me?" I felt all the blood drain out of my cheeks at this gratuitous insult. If I had not known his reputation for upsetting people with his comments, I would have sulked for weeks in a depressed state. By contrast, when Delbrück gave talks at genetics meetings or as an invited speaker at other universities, he was not rude in his comments. All my other encounters with Delbrück after that one insult were friendly, and I was truly happy when he won his Nobel and I sent him a congratulatory letter and received a courteous reply.

## Rudeness as a Lecture Style is Awkward and Breeds an Ill Reputation

The rudest lecturer I ever heard was Stephen Gould. On one memorable occasion in 1987, he was invited to Rice University to give a keynote speech. I happened to be there that week for a dedication of the Julian Huxley collection housed at Rice. Gould arrived a few minutes late because he had to finish a course lecture at Harvard and did not plan slack time for traffic delays. He strode up to the lectern, looked at the glare of lights around him, observed the sweating students fanning themselves with paper, and started off with a complaint about the poor setup and the lack of air-conditioning, his annoyance of having lights shining in his eyes, and his disappointment with the university's trustees for not putting money into nicer accommodations for speakers. He referred to the student union in which he spoke as "the most God-awful architecture" he had seen "for a speaker to face an audience." The British scientists who came for the dedication were horrified by his outburst and one remarked to me that this was probably because he came from New York. I replied that I did not believe this to be the case since I was born and raised in Brooklyn and many New Yorkers have commented on Gould's blunt punch-in-the-nose style of delivery. Despite his belligerent style of lecturing, barbed wit, and caustic comments about critics of evolution, I agree with most of his views on why Darwinism was despised and resisted by the general public as too philosophically radical to make sense. My own brief encounters with Gould have been pleasant and I enjoyed reading his monthly essays in *Natural History*. He was seen by many of my colleagues at Stony Brook and other universities, especially in Great Britain, as beating a dead horse with his stress on a jerky or irregular tempo of evolution which he called punctuated equilibrium. I had fewer quarrels with him over his ideas on evolutionary mechanisms than his overall views on the history of science, which tilted too much toward class-based social bias as the reason for scientific disagreement over how much human behavior is learned and how much is pushed by genetic factors.

## James Watson Enjoys Using Barbed Comments

James Watson can be equally troubling to audiences. He is filled with insulting comments and does not tolerate fools or critics who withhold their motivations. In a 1990 talk addressed to our student Biochemistry Society, he mentioned his involvement with the human genome project and how he tried to keep it as a project for biochemists and molecular biologists: "I didn't want the project to be run by a laboratory that took physicists seriously." He considered physicists as scientists who are "born to hate biologists" and he liked the idea of having graduate students on the human genome project because "they are slave labor and they work on Sundays and they're not old enough to have problems to worry about." Critics have accused Watson of foisting eugenics on the world through the human genome project. He was once asked, "Wouldn't it be terrible to live in a world where your children are all bright and beautiful?" To which he replied, "Not if the alternative is that they're all dumb and ugly." I have always respected Watson's skills and his integrity. While he is as caustic as a scientist can be in expressing his opinions, you can also be assured that he holds nothing back and thus there is no treachery or hidden agenda in his views.

## Graciousness Characterized Linus Pauling's Relations with Students and Colleagues

I found Linus Pauling quite the opposite in personality from Delbrück, Gould, or Watson. He was polite, obliging, and eager to please, and he was always ready to speak to students and encourage them. Gracious would be apt to describe his style. He was portrayed by Watson as flamboyant in his lecture style, although none of the five lectures I heard him give at UCLA or the address he gave to students at Stony Brook when I hosted him, struck me as flamboyant. I did detect a streak of self-importance or vanity when I was chatting with him in 1968 about the progress I was making on my biography of Muller. He said that if anyone ever did his biography, he made it easy for them — he put a date on every page he wrote.

## Barbara McClintock had a Reserved Personality

One of the more unusual personalities in genetics was Barbara McClintock. I knew of her work as a graduate student because it was her analysis of chromosome breaks that provided the model for interpreting radiation sickness. Although Muller and his student Guido Pontecorvo had also worked out the mechanism, McClintock had priority and her name for the process, the breakage-fusion-bridge cycle, was a happy choice and Muller adopted her terminology. I heard McClintock at meetings in the 1950s. She worked with maize (corn) which has a very complex heredity because of the different chromosome sets in the tissues present in a corn kernel. McClintock was also not one to pander to an audience. She wrote for people whom she thought would make an effort to follow her work or look it up if they had not previously encountered it, making her talks virtually opaque to graduate students from other fields of genetics. She also spoke without fanfare and to the point, with one fact or finding heaped on another, often leaving her audience struggling to find a thread that connected all of it.

McClintock was a private person. One summer in 1959 when my wife and I honeymooned at Cold Spring Harbor, I got to hear her comments at weekly seminars and I saw her often in the garden where she worked with her corn, and we exchanged a cheery wave of the hand whenever I passed by. I got to know her personality somewhat better when she called me one day (around 1971) and asked if I would like to take over the reprint collection that Davenport and Demerec, two earlier directors from 1903 to 1959, had amassed. Watson needed the space and rather than throw them in the rubbish, he knew of my interest in the history of genetics and asked McClintock to call me. I made the trip out with an undergraduate student and McClintock was just delightful. "Have a lollipop," she said when I introduced the student. We talked about Muller as well as her work and ideas, which were as radical as ever. She thought of the genetic system as a symphony of floating genes acting as regulators, with the transmission of hereditary information flowing in both directions. Thus, she did not discount the possibility of a theory of acquired inheritance of some sort in which altered proteins could encode genes.

McClintock was appalled by the perception the public had of her after Evelyn Fox Keller's biography A Feeling for the Organism was published. McClintock, now a Nobel laureate, was on the verge of refusing to sign the book when I asked her to autograph it, after she spoke at Stony Brook and joined the Biochemistry Department for a reception. "I shouldn't, you know. It wasn't authorized. And it makes me into a cult figure." In all likelihood, whatever McClintock may have confided about her private beliefs on levitation and other mystical feelings she experienced when meditating, she had not intended these to be either publicized or literal interpretations of her experiences. I enjoyed Keller's book as it was the first clear account of her views and I was impressed at her skill in converting difficult genetic analysis into terms that an educated reader could follow. I was also sympathetic to McClintock's struggle for a job in an era that reserved professorial jobs for men. Despite the prejudice against her, geneticists like Muller had enormous respect for McClintock as a brilliant scientist and peerless observer of chromosomes. I suspect that what she casually told Keller is also true for the astounding model she portrayed to me and my student in her laboratory. It was not something she would have put in print because it was not fleshed out, justified, or tested yet. Instead, it was, as many scientists know all too well, a vision of what might be true rather than a revelation that they are convinced is true. What a noted scientist casually says is not necessarily what he or she may want to see in print.

## Some Scientists Have a Nasty Personality

I consider rudeness a lesser evil as a personality trait than nastiness. I have met a few nasty scientists in my life, most of them not prominent perhaps because their nastiness harms their reputation and limits their talent. I knew one scientist who was hated by the graduate students in his department because he had a reputation for having taken a student's data, used it, and then dumped the student. He also had a reputation for being a sexual bully with female graduate students, who either put up with his terrorizing or quit. I saw him demolish one of his own male students in a qualifying oral examination for the PhD. He also did not get along well with his colleagues, but as a tenured professor, he could not be touched. He was a productive scientist and certainly did good although not trailblazing work. I have long believed that a scientist's personality is largely independent of the productivity or quality of their work. This is not very surprising when one reads biographies of artists, but the public expects this because the most famous artists are more accessible to biographers than are scientists. The general public is not intimidated reading about art but it is frequently intimidated reading about science.

Willard Libby was a chemist at UCLA who is well known for his work on radiocarbon dating and for which he won a Nobel Prize. He served as a commissioner for the Atomic Energy Commission, and I had earlier mentioned Libby's unsuccessful attempts to get my colleagues in the Zoology Department to join him in a commercial fundraising venture. He was an avid supporter of building fallout shelters during the Cold War, which was a source of amusement during one of the frequent fires in the dry season when his fallout shelter was burned and gutted. He also was fair game for bad copy when he provided an interview for the student newspaper advising students that if they wanted to be unhappy and live a lousy life they should go into the arts and humanities, but if they wanted to enjoy life and be happy they should become scientists. He had to write an apology, which was less than convincing, a few weeks later after that remark generated an avalanche of calls and letters of protest to the Chancellor's office.

I had arranged to interview Libby in 1971 for my Muller biography because I had learned from correspondences and independent interviews that it was Libby, as AEC commissioner, who had requested a ban on Muller as a US delegate for the first Atoms for Peace Conference in Geneva in 1956. I ran into him on my way to his office and introduced myself. He asked me who was publishing the biography. After I told him, he said, "Haven't they got anything better to do with their money?" That was not the easiest way to begin an interview with a Nobel laureate and had I not been a student of a Nobel laureate, I might have felt intimidated. The spiciest of his remarks came when I directly asked him why he banned Muller when Muller was the leading critic of Lysenkoism at the time and writing articles about the "death of science in the USSR." Libby exploded, "I don't see why a son of a bitch like that should have any support. He's like so many intellectuals of that time that I knew. They'll take what they can out of their country but they owe it no support. They wouldn't defend their country if it needed them. Muller didn't care for this country at all. He was a spy." Libby added, "He would have done anything to sabotage our defense program. He's no different from that other traitor, Oppenheimer."

Libby was the ultimate Cold Warrior. He saw conspiracy everywhere and no doubt had good reason to be distrustful because, as he told me, "the only person more security conscious than I when we were working with the British on the bomb, was Klaus Fuchs." I have no doubt that Libby saw Muller's subversive side when Muller was a faculty member at Texas, a recruiter for the National Student League, and an editor of an underground newspaper, *The Spark*, that eventually led to his resignation. But Libby was so blinded by Muller's sympathy for Communism in the 1920s and 1930s that he could not believe the sincerity of those who were bitterly disillusioned and became outspoken critics of Stalinism, the tyranny of ideology, and the repression of science and free thought in the USSR during the Cold War years. Although Libby worked behind the scenes to ban Muller, he got along well with his graduate students. One of the superb undergraduates I had in one of my genetics courses did his doctorate work with Libby and found him to be helpful and considerate. He even laughed about the infamous interview in the student paper because he knew that Libby was not an enemy of the arts.

## Ed Lewis was a Warm and Caring Scientist

The most easygoing, friendly, gentle, and modest Nobel laureate I have ever known was Ed Lewis, the Caltech geneticist and founder of the field of pseudoallelism, which is the fruit fly equivalent of gene structure. His ideas on how the structure of complex genes relate to their functions led to a remarkable evolutionary model of how these genes control body segments and give rise to structures like wings, legs, and other body parts. Lewis is my idea of the consummate geneticist. He loved doing his work, did not switch to fashionable fields when molecular genetics was on the ascendancy and Drosophila work was getting eclipsed, and dedicated his working hours to his projects. He also liked to work at night when no one else was around. He was always accessible to my students when they visited Caltech and like Morgan treated them as colleagues. He never complained about his circumstances. I remember how, when I spoke to him on the phone sometime in the early 1960s, he was thrilled that he had just gotten his first office phone and no longer had to run down the hall to answer calls.

## The Independence of Science from the Personality of Scientists in Making Contributions

What has always impressed me about science is its independence from personality. If I were to list on a set of cards the experimental work that led to Nobel prizes and the particular experiments or theoretical principles involved, and on another set of cards I were to list, without their names, the personality traits of the Nobelists who made these discoveries for which they were honored, I doubt that any historian, philosopher, or sociologist of science could match the cards. I did not write a biography of Muller to explain Muller. That would be a fallacy. A scientific biography enriches our sense of the human aspect of the scientist; it does not explain why the scientist chose that field, how the ideas arose, or how the experimental design emerged. Those may be the speculative assumptions we have when beginning a biography, but motivation is murky, often hidden, sometimes denied, and may not even be known to the scientist engaged in the work.

# 13   Science and Politics

We like to think that the freedom of scientific inquiry requires a corresponding freedom in the political structure of the country where science takes place. This is often not the case. Science flourished in Renaissance Italy during a period when most principalities were despotic and the Church had a powerful influence on what could be printed or taught. Major scientific societies were formed under the blessings of absolute monarchs. Censorship was the rule until relatively recently in most countries of the world.

There has never been a free science in the sense that scientists can count on private or public support for research with no restrictions. This constraint is true today even for research in universities, private research institutes, and government agencies. All science has to be funded in some way and few scientists today can accomplish much if they have to support their projects out of their own salaries. The restrictions vary. In reproductive technologies, research on human preimplantation embryos is limited or banned. Certain projects that are too controversial also would not be funded. An example would be projects that are racist or likely to be interpreted as racist, such as studies investigating a link between intelligence and ethnicity, class, or race. A similar lack of funding would be expected if someone were to propose a project that looked for genes predicting criminal inclinations, length of prison sentence, or type of crime for which the conviction occurred. Any blatant or covert anti-Semitic research project would also be unfunded.

## Research on the Nature-Nurture Controversy is Difficult to do Without Any Bias

Research on the relation of genetics to behavior is not as simple as it sounds. Clearly someone with a political ideology, say a sympathizer of Nazi ideology who wanted to prove the inferiority of all ethnic and racial types, except Teutons and their descendants, would not be deserving of government grant support. But if someone claimed without evidence of conscious bias that all they wanted to do was to see if behaviors such as aggressiveness, shyness, inquisitiveness, generosity, conservatism, or caring are primarily inborn or acquired, such research also might not be as easy to fund as one might think. Many scientists, not just those deemed to be ideologues of the left, might have concerns about such a project. Indeed, it raises several issues. One is the legitimacy of doing controversial research without due effort to rule out unconscious bias in the design of instruments measuring genetic or environmental influences on traits, which are susceptible to social influence. Unlike most other types of research, research dealing with the nature-nurture debate is often used to shape national policies and can, intentionally or not, damage individuals who belong to classes, races, or other categories looked upon by those who read accounts of the research as collectively inferior in some major way. This outcome might not be the intent of the scientific investigators, but the history of abuse of such research throughout the twentieth century does make it appropriate for government agencies to consider carefully whether to fund such research.

I remember the bitterness of such debates in the 1970s when those who were doing psychological research on intelligence testing suddenly found themselves under attack. I found it strange that genetic claims on such differences, especially between races, were being made not by geneticists but by educational psychologists. Their approaches were very different from the experimental genetics approach of isolating and mapping individual genes and eventually identifying the functions of each of these genes. Instead their approach was statistical, based on correlation studies, and clouded by uncertainties about the subjects used and whether they were influenced by a number of factors for which controls were lacking or impossible to obtain. Twin studies, adoption studies, and family relatedness studies are limited by many of these poorly explored circumstances. To what degree do agencies try to match homes for separated twins? Do they seek relatives first before trying total strangers? How long were the twins together before they were separated? Did the separated twins ever meet in their formative years? Did the adopting parents know of the other twin's existence?

I have never been a fan of this approach to human genetics. Many of my colleagues consider this "soft science" because it resembles studies of heredity in the nineteenth century that did not have the tools of genetic analysis, gene mapping, and similar means of associating specific functions with particular genes. I do not doubt that within a generation there will be genetic analyses of individual genes that act on personality traits, memory, learning, and other basic behaviors. In all likelihood, there will be thousands of such genes and a great deal about the nervous system will begin to make sense at a molecular level. Right now, that understanding is close to zero, almost like studying genetics in 1897.

## Scientists are Divided on the Reliability of Behavioral Studies Implying Innate Differences

Some of my colleagues would consider my opinion wrong as well as unkind. They would argue that a great deal can be learned from statistical studies of population differences and that it is not necessary to map every gene and isolate its function to see that there are genetic components making populations differ on socially significant traits. They stress their sincerity of lacking bias and just being curious about an issue that can be looked at from a scientific approach. They see nothing wrong with using twin studies or other standard human genetic approaches to identify these genetic influences, if they do indeed exist. They claim that the correlations are sound and that such genetic differences are real, and they argue that the political motivation of liberalism or leftist leanings or the ideology of environmentalism motivates their critics. Those who are critics of such research make similar pleas of neutrality or objectivity and claim that conservatism, elitism, or an ideology of biological determinism motivates those who enter this kind of research. Even if each side is correct in saying that they are trying to be both objective and sincere, both sides may underestimate their own hidden biases when studying socially significant traits.

## Governments Sometimes Use Science to Justify Bias or Ideology

Unlike this ongoing dispute in a democracy that tolerates divergent views (even if it does not correspond with research), there is plenty of evidence that governments on both the left and the right have manipulated science to support their own ideologies. No one who reads what happened in Germany after Hitler became its Chancellor and brought Nazi ideology into government policy can doubt the damage done to science by such political interference. Scientists who were Jews, socialists, or anti-Nazi in their political leanings were purged from universities and research institutes. Fields such as anthropology and human genetics were converted into instruments of propaganda to support Nazi ideology. Medicine, psychology, and the social sciences were also heavily influenced by the new government demands. Scientists who were freest during these thirteen years of Nazi domination were those whose work had no social relevance. But once the war broke out, little basic research was supported by either side, censorship limited what could be published or discussed at meetings, and most scientists on both sides felt a duty out of patriotism to help with the war effort.

A similar stamp of ideology occurred in the USSR, especially in the years after Stalin secured his hold on the Soviet Union. A gradual purging of scientists began in 1936 after some initial skirmishes over a growing debate in both agricultural genetics and human genetics. One skirmish was over the upcoming International Congress of Genetics that was scheduled for 1937 in Moscow. There are five-year intervals between such meetings and the 1932 Congress was held in Ithaca. As the planning began in earnest in 1933, demands were made on the international committees to keep German geneticists out. Internally, debates among the Soviet planning committee centered on the possible banning of any human genetics at these meetings because they were tainted with eugenics or racism. Throughout the 1920s, there was a range of views on Soviet genetics. Some were interested in eugenics whereas others opposed all eugenics as spurious. Some established a research center for the study of human genetics as a science (no eugenics was preached or allowed into the publications), but the nature-nurture controversy was actively studied with many sets of identical twins who were put through batteries of physical and mental tests.

# The Lysenko Controversy Demonstrated the Vulnerability of Science to Government Policy

Much more serious than the growing impasse over who would attend the 1937 Congress of Genetics was the movement in Odessa that was started by Trofim D. Lysenko, a plant agronomist who switched to heredity. His studies of heredity were inspired by the Russian equivalent of Luther Burbank, I. V. Michurin. Lysenko called his movement Michurinism and the West called it Lysenkoism. To most western geneticists, it was a revival of Lamarckism. Muller, who was in the Soviet Union from late 1932 to early 1937, witnessed the shift from a modestly free society where debates were open and free-wheeling to a closed society where strongly supported experiments and principles of Mendelian genetics were denounced as bourgeois or fascist. Two of Muller's students who had come from the USSR to study with him at Texas were arrested and executed as Trotskyites. Lysenko at this time had the secret backing of the Party and he arranged for a gradual takeover of the leadership of Soviet agriculture and its main research programs. At a meeting in December 1936, Muller and the Mendelians confronted Lysenko and his supporters during a scheduled debate in Moscow. The debate was acrimonious with hostile shouts and boos by each side over the perceived rigidities, ideologies, and errors of their opponents. Muller left the USSR shortly after the debate, realizing that future genetic research was impossible in the Soviet Union under Stalinism. His infatuation with Soviet communism had ended.

After the war, the debate resumed. In 1947, Muller broke his silence as President of the International Congress of Genetics held in Stockholm, gave a history of the Lysenko controversy, and named his colleagues there who had been killed, arrested, or banished from the field of genetics. The Soviet delegation protested and left the Congress. A sham meeting was held the following year to debate the issue in Moscow. Lysenko sprang his trump card and announced that his views had been endorsed by the Party. In one of the saddest moments in the history of science, the leading Soviet geneticists got up one by one to recant and announce their conversion to Michurinism. For nearly 20 more years, Lysenko held sway and his supporters became professors, editors, and institute heads throughout the Soviet Union. Lysenko argued that heredity was shaped by the environment — whatever heredity that was in a plant or animal could be shattered by the environment and the organism could be retrained by an appropriate environment to inherit what was desired. Lysenko's theory was appealing to a society that sought quick environmental change and wanted to believe that both human behavior and plant biology could be changed radically with good will. Lysenko claimed that he could change winter wheat into spring wheat or rye into wheat. Lysenko played on people's fears of fascist ideology and the upper-class prejudices that fed the eugenics movement in the United States. He and his supporters argued that Mendelism was a bourgeois science invented by a priest and thus was not to be trusted. He argued that the theory of the gene was deliberately used to justify a creeping change that blocked progress in all living things, including the capacity for children to benefit from education and "progressive" geneticists to alter Soviet agriculture and produce abundance in all growing seasons. Lysenko failed for many reasons. He could not deliver the goods he promised and yields of grain were no different under his system than in the past. He claimed sabotage by "wreckers," but that claim could only work for so long. It also tainted Russian science and other scientists shunned Lysenko as a fraud, many of whom even blocked his efforts to elect his supporters to the Soviet Academy of Sciences.

Even after the collapse of the Soviet Union, genetics in Russia and its allied republics has yet to recover fully from a generation of destruction by Lysenko and his supporters. Communists around the world during those debates from 1948 on faced big difficulties. Some of their strongest supporters were scientists but because they lived in Western countries, they knew that genetics was a legitimate science as they saw the advances of gene theory and its immense success in molecular biology. However, they dragged their feet following an international movement for Lysenkoism. There were many scientists, not just geneticists, who soured on communism after witnessing these attacks on science.

## Cold War Pessimism Misled or Intimidated Scientific Studies

Democracies are not immune from either government or political interference. During the Cold War, mutual paranoia prevailed among the advisors of the United States and the USSR. Each side saw the other as attempting to destroy it, sought a defense that would serve as a deterrent, and made patriotism almost synonymous with acceptance of government policy on sensitive issues. For the United States, a major issue was its nuclear defense capabilities. For reasons that in retrospect are difficult to understand, advisors in the 1940s and 1950s interpreted concern over radiation damage as communist propaganda that was intended to wreck the US's development, testing, and deployment of nuclear weapons. It did not matter if the concern was over abuses of radiation in medical practice, industry, or the weapons testing program. All concern over low doses of radiation as a source of mutations or cancers in the population were either denied or described as propaganda. These views were echoed by scientists themselves, creating a split between those in the life sciences who recognized the vulnerability of genes and chromosomes to radiation damage and those in the physical sciences who scoffed at low dose effects. It did not matter that children's milk teeth showed the presence of Strontium-90 from the milk they drank, which was obtained by cattle grazing on fallout-tainted fodder. It also did not matter that the bones of animals grazing in Utah and Nevada had bone sections that were so radioactive they could expose x-ray film. And it did not matter that communities in the Western states and the troops sent there for military maneuvers were put at risk. The official policy was one of public denial and secrecy.

One of the sadder front-page items I read in the New York Times was an account of a public disinformation policy that was recommended by the Eisenhower administration to confuse the public on radiation safety through contradictory stories about radiation. In the 1950s, it was not unusual to read claims that radiation in small doses was either harmless or actually beneficial because it increased hybrid vigor. I heard Edward Teller give a talk at UCLA in which he mocked concern over radiation, standing up from his chair and saying, "There, I have exposed myself to as much increase in radiation by raising my elevation as you will get from worldwide fallout." He went on to criticize biologists "who should know better. After all, radiation is good for us; it speeds up evolution and induces beneficial mutations." It is not uncommon that scientists appear to be ignoramuses to each other when they are in different fields. Teller was not well informed about evolution because he would have otherwise known that damaged genes (spontaneous or induced) far outnumber beneficial mutations by a factor of 100 to 1, and even if there were beneficial genes induced by radiation, these would only be of value if there was natural or artificial selection to prevent the reproduction of the other 99% of individuals receiving harmful genes.

Although such political influences may have little effect on the validity of the work that scientists do, they are still vulnerable to them. During the 1870s and 1880s, Robert Koch in Germany and Louis Pasteur in France became bitter enemies. Both were nationalistic patriots and Pasteur so loathed Germans for their humiliating victory over France in the Franco-Prussian War that he turned down honors from the Prussian government later in life and dedicated several of his papers to revenge and the future defeat of Germany. He rejected Koch's analysis of anthrax with as much passion as Koch rejected Pasteur's. They heaped polemical articles on each other in an era when editors did not block personal invective in publications.

## How Nationalism can Shift Interpretations of Science

It does not take a war to bring out nationalistic feelings. When I edited a collection of articles on gene theory for a paperback book for Dickenson Press some years ago, the editor of the publishing company told me that my book would be translated for a German edition. I received a copy of that edition and to my surprise, they had removed an article by G. Pontecorvo and replaced it with an article by H. G. Wittmann. It was not that German names and German contributors were altogether lacking. I presume that the German editor at Gustav Fischer Verlag felt that the book would be more marketable if the German contribution to gene theory was augmented.

A similar experience was related to me by Charlotte Auerbach when I visited her in Edinburgh. She had produced a popular book on genetics and the atomic age, and a German edition was translated and a copy sent to her. The book was edited with so many changes suiting the German editor's political views (himself a prominent scientist) of what radiation ought to have done to the population that she felt ashamed for it to be published under her name. Despite her protests, the publisher stood firm and would not publish a retraction or insert her views as a supplement.

Nationalism is not unusual as a motivator for scientists. They like to think of science as serving the people. A few may have a universal view of that service, but many others feel quite happy with that benefit being assigned specifically or only to one's country. Such nationalistic fervor is not a new trend. I still cringe when I read Leonardo Da Vinci's obsequious letters to his patrons trying to convince them that his military engineering skills would benefit them and defeat their enemies. Richard Goldschmidt, who fled Germany as a Jew when Hitler came to power, still seethed with rage over the injustice meted out to Germany by the Versailles treaty. His rage was as genuine and tinged with German nationalism as was Pasteur's French nationalism in response to the equally humiliating peace treaty imposed by Germany on France after the French military collapse that ended the Franco-Prussian War.

Yet if one were to ask each of these nationalism-inspired scientists what they considered the virtue of science to be, they would be virtually unanimous in praising it for its universality. Scientists believe that what they find and interpret is purged of personal and national bias and represents something that anyone anywhere can verify. Even those who were patriotic as citizens during wartime quickly return to this image of universality at the end of the war. When British geneticist William Bateson, whose son was killed in battle shortly before the Armistice, was invited to attend an international meeting shortly after the war ended, he deplored the attempts to bar German scientists from attending and argued that such nationalistic prejudice had no place in scientific thinking.

## Science Denial can Eclipse Evidence and Reason

The presidency of Donald Trump was marred by numerous examples of science denial by his and his appointees' positions on two major issues. The first was their denial of climate change associated with the use of fossil fuels. The Trump administration rejected the validity of science-citing critics who claimed that climate change was man-made. Instead, Trump's supporters claimed the changes in weather were just consequences of solar changes or atmospheric perturbations long associated with Interpretations of the ice ages. The only "good" alternative to fossil fuels, they argued, was nuclear energy.

The second failure of the Trump administration was how it handled the COVID-19 pandemic of 2020. In his public appearances, Trump tried to reassure the public that this was no worse than a bad cold and that it would end as soon as the summer came. He also erred in urging an end to "lockdown" or self-quarantine movements by various states, or allowing that decision to be made by the 50 state governors or their legislatures rather than through federal action. The Trump administration also failed to provide the necessary effort for national calamities (e.g., FEMA for hurricanes and floods) and shift medical equipment so that those with surpluses could help those without protective masks, clothing, and gear. Either Trump or his designees ignored the advice of the National Institutes of Health or the Center for Disease Control, leading to the United States having the worst record in number of cases and prevention of rebound infections from the premature reopening of business. A third area of collision of the Trump Administration was Trump's views on religion. Trump's supporters included a sizeable number who reject evolution by natural selection or any scientific interpretation that is inconsistent with a fundamentalist view of the events of Genesis (the first seven days of creation attributed to God). Trump has also supported the political agenda of the "religious right wing," which includes federal financial support for religious school education and the right to fire or not employ persons who advocate birth control, elective abortion, or LGBT rights. From the viewpoint of Trump's critics, he favors the right to discriminate on religious grounds.

## Science can Sometimes Flourish Under Despotic Governments

Although tyrannies that dictate how science should serve their ideology tend to fare poorly in the basic sciences, they can be very effective in applied science. The Soviet Union is a case in point as they beat the United States in the first salvo of the space competition with their launch of the Sputnik in 1957. And if Willard Libby was correct when I interviewed him, the Soviets were the first to explode a hydrogen bomb (ours, he pointed out, was a device, not a bomb). But I am less concerned about such tyrannies because of their relatively short duration before they are defeated or shift policy.

Science is more likely to benefit or be harmed from the way it is organized in different countries. In both pre- and post-Soviet science, leadership and power came from its Academy of Sciences, which was largely composed of old men, many past their creative prime. At Germany's prime, superstar professors headed departments and if such professors were at their peak when recruited, they had many productive years to develop their skills. Germany treated its professors with such public esteem and financial reward that its universities were the envy of the academic world. The United States probably owes its world domination in the sciences to the large number of first-rate universities it has, which places together several first-rate scientists who can stimulate each other's work in a single department, and at least after World War II, the shift of funding to competitive, peer-reviewed, government grants for almost all basic research. While such happy arrangements make science flourish, there is no guarantee that the United States, or any government for that matter, will continue to support its scientists and provide optimal conditions for their work. Governments serve the people, and those who elect governments frequently shift their priorities as many segments of society compete for limited resources. Science, like all other segments of society, must learn to live in lean years as well as generous ones.

# 14 Science and the Arts and Humanities

An uneasy relation has existed between science (including its applied forms in engineering, medicine, and mathematics) and an area of the liberal arts known as the arts and humanities. About fifty years ago, the conflict between the two ways of looking at the universe was almost like open warfare. C. P. Snow, a scientist and novelist, described the skirmishes in his classic *The Two Cultures*. Snow, whose background was in the physical sciences, argued that scientists tried to read widely, go to museums, listen to classical and contemporary music, and recognize the value of the liberal arts in their fullness. By contrast, those in the arts and humanities feared science, shunned science, or looked upon science as an enemy of the arts and humanities. He considered the reasons for this quite different response of the two camps and felt that it was largely a matter of scientific illiteracy and that like chicken soup for a cold, a little bit of the second law of thermodynamics might clear the artistic sinuses.

I am skeptical of such assessments. I know quite a few of my colleagues in the arts and humanities who enjoy reading about science. They are science buffs and like to go through the *Science Tuesday* section in the *New York Times* to enlarge their sense of the universe. Even if they do not understand science beyond their high school education, they look upon it for stimulus. I once sat with my colleague Terry Netter, a fine modern artist and former Director of Stony Brook University's superb Staller Center for the Arts, and Paul Grannis, a physicist at Stony Brook who was in charge of the team that detected the "top quark." Netter told Grannis how much he loved the names that physicists gave to their arcane chunks of the universe, from black holes to quarks to naked singularities. "Could you tell me what a naked singularity is?" he asked and apologized that he might not even understand the explanation, despite the fact that he had just finished two huge oil paintings that he called *Naked Singularity I* and *Naked Singularity II*.

## Interests in Science and the Humanities Vary Across Individuals

I know that many of my colleagues in the sciences attend concerts at the University and many of them play musical instruments. Some scientists were quite proficient. E. B. Wilson was well known for performing in chamber music groups as a cellist. Edward Teller was adept as a pianist. Many of the biochemistry, biology, and mathematics majors I know are participants in university orchestras or sing in chorale groups or campus operas. I was impressed at the skilled artistry of Fritjiof Sjöstrand, a colleague at UCLA and an electron microscopist of international repute, whose muted colors in his landscapes reflected the transition between realism and geometric form, not in the glimmering sharpness of Sheeler's industrial world but in the subdued vision of an overcast subarctic world.

The conflict is real, however, among many scientists and many artists and writers. I do know some scientists who are tone deaf, cannot appreciate music no matter how hard they try, and consider a trip to a concert as torture. Muller and his wife, Thea, were both musically bereft of talent and on the occasions when Muller did attend command performances, such as the Bolshoi ballet in Moscow, he would more often be leaning over to engage some student in planning experiments or discussing results. While Muller's appreciation for music was near zero, he enjoyed looking at art, occasionally tried his hand at poetry, and read extensively. I have rarely encountered a scientist with soul so dead that nothing in the arts or humanities could penetrate. But I have known scientists who temporarily became insensitive to the arts and humanities. One of the most chilling descriptions of this can be found in the short autobiography that Darwin wrote for his family in his middle age. He described how he read Shakespeare with voracity as a youth and could quote passages at length, but after he began to immerse himself in the writing of his scientific books and amassing the evidence to support his theory of natural selection, all of the world of humanities — music, literature, plays, opera — took on such an artificial form that he abandoned efforts to revive his appreciation of the arts and humanities, considering them cauterized by a full-time commitment to science. We know that he mellowed out later in his life and was again reading novels aloud to his wife.

I had a similar experience when I was a graduate student working fourteen-hour days gathering flies, examining them, mating them up, and reading the scientific literature on my dissertation research extensively. As I walked back to my apartment, everything was Drosophilized. I saw pupae and larvae in the clouds and in the cracks of sidewalks, and I heard the churning of larvae in the background sound of the streets late at night or early at dawn. I recall trying to read a novel and putting it aside in disgust at its artifice. The real world I was in was so overpowering that everything else seemed like a distracting contrivance. I realized that I had become Darwinized! I cured myself of my folly by reading Stendhal's *The Red and the Black* and *The Charterhouse of Parma*, books I had bought when I first came to Bloomington but had not yet read. It took about 75 pages of forced reading before the novel opened up a world that I could once again enter.

Darwin's loss of aesthetic sensitivity was not unique to scientists, however. In Tolstoy's *Anna Karenina*, the hero, Levin, describes an episode in his life where he moved from the country to the city. Everything struck him as artificial and distasteful. He was horrified as he sat through his first theatrical performance and saw the dinners, balls, swirl of cultural life, and conversations as a monstrous lie that ignored the reality of the simpler values and direct experience of country life. Yet after a few weeks of total immersion in this urban atmosphere, he was transformed and became one of them!

# C. P. Snow Drew International Attention to the Culture Wars

The public perception of science often reflects the "two cultures" view that Snow put forward. There is a fear of science as a way of seeing the universe that leads to emotional death. The scientist is seen as possessing reason devoid of feeling, passion, romance, or values. As pure reason, science is seen as a threat to humanity. Pure reason obliterates the heart, and the world is the loser. Pure reason gives us the munitions makers, the Wannsee conference for the Final Solution, and Los Alamos and the atomic bomb. It is also seen as giving us domino theories, détente, containment, Cold Wars, pollution, "our friend the atom," planned obsolescence, and an endless flood of gadgets and items we feel compelled to buy.

Equally exaggerated is the perception by scientists of the scientifically illiterate. They are deplored for living in a world where mysticism, superstition, pseudoscience, wishful thinking, fantasy, denial, ignorance, fatalism, and a mind disembodied from the brain flourish. Such a world is horrifying to scientists as it allows racism, bigotry, self-deceived paranoids, quackery, charlatans, and tyrants to trick and exploit the gullible, foolish, uninformed, and desperate.

Both views are distortions, of course, although it is easy sometimes to lapse into extremist thinking about non-colleagues who come from the opposite camp . A good liberal arts education should include exposure to the arts, humanities, and sciences. Reason and the emotions are both part of our heritage and our humanity. All humans must constantly apply their skills and knowledge in a moral universe of contending values. A science course that does not reflect on its significance and connection to our humanity fails in the same way that an art history or literature course fails if it stresses plots, names, events, and dates but does not put things within a broader context of the human experience.

## Many Novels do an Excellent Job of Assimilating Science

There are artists who understand the "two cultures" issue and they handle it well without destroying their integrity as artists. I consider Turgenev's *Fathers and Sons* a superb example of that tension between the old and the new, the emerging world of science and the familiar world of family and spiritual values. We learn of the father's world when the son returns home with his college friend Bezarov. In particular, we see that the comfortably settled middle-aged parents know how to live in a world that may be unjust but which doles out enough to those whose good fortune and talents permit them to enjoy it without exploiting it. Bezarov, however, does not see this as worthy of a new generation. He brings less of a revolutionary vision of a utopia to replace a flawed world than a cynicism that rejects everything, and Turgenev prescribes a philosophy to him — nihilism.

Bezarov is a product of the mid-nineteenth century revolution that replaced natural theology and a God-saturated world with a rational world that is helped along by the practicality of the industrial revolution, whose science and technology kept God out of daily conversations except for dinner time formalisms and obligate Sunday services. The world of reason descended on medicine, industry, universities, and science wherever it was studied. Out went the biases of theology and the parochial values that contending religions pushed on all fields of knowledge. Bezarov's idol is Claude Bernard and the new physiology that sought to work out how the body's organs functioned. From this new perspective, man is a machine, wonderful to behold and even more exciting to dissect, analyze, and understand. Bezarov's reductionism acquired during his medical studies gave him a new power that medicine quickly recognized as its future. The world of surgery, accurate diagnosis, differential pathology, and a scientific explanation for all illnesses would give humanity enormous power to defeat the intrusion of death while we still have years to go in our natural life cycle.

Turgenev recognizes the limits of both science and nihilism. Science without heart and values becomes cold, and its Midas touch turns everything into an empty materialism. Nihilism is intrinsically doomed to fail because its only function is in devouring error and leaving a wake of disillusionment. Bezarov becomes a victim of his own limits. The germ theory had not yet emerged and he nicks himself with a scalpel and dies of septicemia. Medicine had not yet achieved perfection in its control of diseases and neither has it reached that perfection today despite our panoply of antibiotics, organ transplants, and avalanche of new technologies and options. Even worse, we are shocked that a coronavirus pandemic in 2020 can revive memories of the flu epidemic of 1919, the London bubonic plague of 1665, and the bubonic plague that wiped out one third of Europe's population in the 1350s. Nor will medicine reach that perfection when the human genome project is fully before us, every gene available in its sequence along the chromosome, mapped and available for us to clone and to use. There will still be suicides, premature deaths, chronic illnesses, accidents, epidemics, and crummy combinations of genes that will not be responsive to therapy or whose damage will not be undone by replacement of a gene product or substitution of a mutant gene by a normal gene.

Thomas Mann in *The Magic Mountain* also does a good job with science. His hero, Hans Castorp, becomes tubercular, and in the first decades of the twentieth century there were only rest cures and surgically imposed relaxation of one lung. Fresh air, good food, and rest restored some while others had their galloping consumption eat away at the infected vital organs, with little that could be done to keep death at bay. Mann gives us a portrait of Europe just before it immersed itself in the First World War. We are introduced to the clash between the old religious values and the new socialist materialist values in the debates of Naphta and Setembrini. They lean on each other like the Cold Warriors to come, each needing the other as the ultimate evil to resist. Mann loves to use them for his own reflections on the mood of the world — a troubled one — and he is full of justified foreboding. We see his admiration for science in his symbolic treatment of a lung x-ray that Castorp treasures of his tubercular girlfriend. It is her inner self that the x-ray reveals. The artist often takes science and studies it like a prism that reflects light, finding connections to the imagination that no scientist can permit in the realm of the scientist's material world. It does not matter that an x-ray is not a soul or the essence of Castorp's girlfriend. Neither does it matter that Netter's Naked Singularity I or II resembles some ylem or proto-matter or whatever a physicist might consider the state of an incipient universe in a void to be.

We return again to the real world at the end of *The Magic Mountain* where science is no longer immersed in an ethereal universe of the artist's imagination, where a description of the early stage of embryogenesis and organogenesis from the fertilized egg to an emergent being transports us with awe. Instead we are shattered by the image of Castorp the good citizen and engineer, no longer the patient confronting the death his disease might bring but the frontline soldier hurtling past barbed wire, his monstrous iron and steel ships heaving explosive missiles miles away to shatter fortifications and devour life in the wild Satanic orgy of war that reason and its engineers, willing servants, and elected governments with its holiest of virtues — patriotism — dictate.

At the same time, Mann can describe snow for thirty straight pages without repeating himself, revealing his command of visual imagery, the sensual experience of being engulfed by snow, the swirling eddies, the myriad of flakes, the exhilaration of being on skis, the panic of being lost or immersed in a white out, and the triumph of finding one's way home. It is a description worthy of the most meticulous dissection of a barnacle by Darwin, but far more elegant in the richness with which it penetrates our emotions and our experience of the fright, isolation, and triumph of the skier.

# Does the Spread of Science Lead to a Bland Universe?

Years ago, when psychological conditioning inspired by Pavlov's dogs had penetrated every psychology department, I remember assigning one of my classes *Walden II* , B. F. Skinner's utopian novel of the good society. Unlike *Brave New World* and other pessimistic views of what life can be in a technological universe, which seems to dominate the literary creativity of twentieth century fiction writers, Skinner offers a world that works, at least for him. It is a world where conditioning determines our behavior, and the hurtful things we do are substituted by rewards or punished out of our lives by the psychological devices that milk us of our foolishness. In its place we are offered an emotionally flat universe. Tempers rarely reach incandescence, love is muted passion, one reads to inform the mind of useful information, and all that irritates it — history, the supernatural, dysfunctional families, the stuff of novels past — is trashed and a bland non-fiction prevails. People are polite and boring. Utilitarian arts prevail — cabinet making, pottery, traditional quilting. It is a world where hobbies occupy the day and the world recedes into insignificance.

The conflict of the two cultures has largely subsided. There is still a lingering fear in the non-scientist's mind that science has gone too far. It emerges whenever a new technique or principle that has human applications is introduced. Prenatal diagnosis, donor insemination, in vitro fertilization, DNA diagnosis, gene replacement therapy, the human genome project, DNA chips, artificial organs, organ transplants, the world wide web, space stations, neural tissue transplants, and dozens yet to come will shake the belief of some that their humanity is at risk and that science is about to reduce it to tinker toy components. It is a story as old as science and has inspired legends such as that of Icarus falling to the earth as his feathers detach in the warm sun. It has inspired the Faust legend of the scientist who sells his soul for knowledge and the novelty of unending discovery, and the fiction of the scientist gone amok, Dr. Frankenstein, who creates monsters who return to life only to harm us. In modern form, we see it in the emotionally empty physicist who gives us "ice 9" in Vonnegut's *Cat's Cradle*. I do not see an end to this fear of science, and it is not irrational. While science may not create hate, bigotry, war, ideology, and envy, it certainly lends its support to governments, regardless of the justifications for war. Nor can we demand of science a rational analysis and sophisticated program that can eliminate all the insecurities, delusions, desires for power, and personal convictions of righteousness that lead the few to captivate the many and engage them in war. Science is seen as a threat less for what it instigates than for what it serves. The Da Vincis of the world unite in their respective countries and convince themselves and their leaders that they will defend its citizens while destroying all their enemies.

Fortunately not all fantasies can be realized. There are limits that the real world imposes on our imaginations. We cannot fly back in time, live forever, or become gods. We are limited by our genes, the stuff our bodies are made from, the properties of the 90-odd elements that compose all matter, and the laws that our material universe obeys. The artist has no such limits. We can create jolly green giants who tower forty feet tall with the same proportions as living men, add wings to our heroes and make them fly, and have a Superman who flies by the mere exertion of his muscles. In our visions of hell and the world of gods and demons, we can imagine human-headed beasts and female hydras with snakes as hair. We can also imagine growing backward in time until we dissolve into a zygote, people communing through mental telepathy, or having a mass pray-in for all guns in the world to turn into melted butter. We can believe that our soul has memories of past lives that recur in unpredictable moments and reassure us of our immortality and our eventual union with a godhead. All of this can be imagined, but belonging to a realm outside science, none of it can be realized in the world of matter. For the general public, that world may be proclaimed to be as real as the material one, but few truly believe this in their private thoughts. They fear death; they do not welcome it. They work to earn a living; they do not pray for money to miraculously appear in their purses and wallets. They know that churches with worshippers in it can be leveled by floods and hurricanes. They know that Holocausts can come and the innocent can perish for no other reason than that they were born. However the world of imagination may work in our fantasies, it does not work in our daily experiences. The scientist is forever sobered by reality. Facts may destroy theories and force their modification or replacement. Theories may be the magnificent imaginations or constructions of talented minds, but they become nothing when experimentation proves them wrong. As E. B. Wilson so astutely realized at the beginning of the twentieth century, science cannot detach itself from the "disciplined imagination."

# 15 Science and the Social Disciplines

It is a truism that everything we do is shaped by our environment. I know of no scientist who is oblivious to that fact. But exactly what is the relation of society to the sciences? It certainly determines what is applied, what is sponsored by state or private funding, when certain science can be done, and when it is forbidden. Regulation is not only expected but necessary for most applied science. I would not want to be at the mercy of market values for new prescription drugs and run the risk of another thalidomide-type disaster where an unregulated German market in the early 1960s produced some 8,000 armless and otherwise deformed babies. I mentally send my thanks to Frances Kelsey at the Food and Drug Administration for her staunch protection of the American people at a time when she was under attack from her drug company's critics as "an officious old bureaucrat." I am also aware that there were responsible drug companies in the US that rejected thalidomide as a worthless tranquilizer, but the integrity of some companies is not enough to protect the public.

## Social Scientists see Science as a Construction of Reality

In sifting through what society does in shaping our values and the way we work and live our daily lives, I recognize that societal influence is enormously pervasive. I owe my American values to socialization from the thousands of pledges made to the flag in my school years, which helped to make me who I am and created an image of America that I cherish. I am not as persuaded that this social influence in a direct way makes me "construct" a reality that I call science and equate as true, real, proven, or factual. I know that some historians, philosophers, and sociologists believe that I am deluding myself if I think that they are really real. They tell me this in front of my own students! Like most of my fellow scientists, I believe that there is a difference between something demonstrated or proven and something that serves as a hypothesis or model to guide my thinking. It is the great ability of science to test things and show whether predictions are borne out or not. The fact that science works in predictable ways and that pseudoscience and other claims to knowledge do not have a very good or consistent track record assures me, as it should the general public, that there is something special about science that makes it work. Virtually all scientists will tell you that this "something special" is the use of reason to look for consistency and accuracy, design ideas or testable hypotheses, and test things and look for agreement or inconsistency with predictions.

## Some Scientists Deserve Recognition as Heroic Figures

While this ongoing debate over the nature of reality is part of today's issues that engage both scientists and their critics, there are other aspects of the influence of society on science that I think are worthy of our attention. I can think of some beneficial ways that society relates to science and some not so nice ways. Of course, the reverse is true because science can have beneficial or perverse effects on society. I will begin with a positive instance — germ theory. I like to argue that the most important people of the nineteenth century were not Lincoln, Marx, Napoleon, or Queen Victoria but Semmelweiss, Pasteur, Lister, and Koch. Those political leaders certainly made profound contributions to the lives of others (not always to the good of all), but I could and do argue that the four scientists who gave us the germ theory have saved hundreds of millions of lives, led humanity in the twentieth century to adopt family planning, and given us the world we now expect — low infant mortality rate, relatively high mean life expectancy, the expectation that the children planned and born will grow up and become parents, and that our own lives and theirs will largely be spared an early death from child bed fever, tuberculosis, typhoid fever, diphtheria, dysentery, and pneumonia. Those illnesses were the great killers of the past, and deaths from infectious diseases until only a century ago killed far more people than any other activity. We worry about wars, but all recorded wars have killed only three percent of humanity. All of the early pioneers in the development of germ theory were ardent in their desire to make their colleagues, the public, and their governments aware of the value of public health measures to reduce the incidence of infectious diseases.

## Scientists Can Also be Villains in Their Applications of Science to Society

Quite the opposite effect on society is the role that science played in the implementation of compulsory sterilization, beginning in 1907 in the State of Indiana, as the way to diminish what were once called "the unfit" — mostly people troubled by insanity, mental retardation, pauperism, and unsavory criminal records. This movement was based on several fallacies. In the 1880s, many critics of the unfit believed that they were fixed in their hereditary traits and that social failure was largely a genetic matter. They believed in the virtually guaranteed transmission from parent to child of the bad trait, which they thought could be eliminated after no more than three genera-tions of vigorous application of sterilization to unfit adults of reproductive age. All of this was false, but they did not know it. To our horror, we find that the sterilization movement was driven not by some crazed ideologues representing robber barons, bigots, or middle-class exploiters, but rather by physicians, social reformers, advocates of public hygiene, and many of the leading intellectuals of the day. The eugenics movement was absorbed by Nazi Germany and the horrors of its thoroughness in expunging those it considered unfit still echo today. Eugenics continues to be justly feared for its potential for abuse. It is important to remember that eugenics was not invented by Nazi ideology but rather incorporated by them from an American concept dating back to the 1880s. British eugenics, founded by Francis Galton at about the same time, was much more benign and sought to increase the reproduction of the most talented of its men and women. Unlike the American movement, the Galtonian ideal of eugenics was seen as a moral duty or a secular religious feeling. Why America sought a legal path to eugenic change whereas Great Britain sought a moral path is something that social historians and sociologists may eventually determine. It tells me that what science does when it is applied to society as a whole can work in different directions; to a degree, it may be more influential than many like to admit, and these outcomes are made through the efforts of a few committed individuals who convinced their intended audience, whether that audience consists of legislators or fellow intellectuals. For America, a major voice was that of a physician, Dr. Harry Clay Sharp, whose reports to the Governors of Indiana and insistent requests to legislators that the State pass a law to sterilize the unfit took more than five years of lobbying before it won success. For Great Britain, it was the books and articles that Galton published that drove his persuasive efforts (with little success) to make his peers have more children.

## Why do Scientific Experts Sometimes Disagree on Controversial Issues?

The public has good reason to be puzzled over another aspect of science and society. Experts frequently disagree, and this may be seen in court cases where the defense and prosecution hire experts to testify to juries about the validity of evidence. It is also seen in public hearings when testimony is obtained on safety issues for building such things as nuclear plants, dams, malls, highways, water treatment plants, garbage dumps, factories, or other novelties that may alarm local residents. One could argue that if science gets us answers, why then do we have disagreement among scientists who testify? What most of the public does not know is that much of applied science is not clear cut. There are many complicated consequences of applying science. Scientists (engineers primarily) may know with absolute certainty that a nuclear plant can never explode like an atomic bomb. This is correct. But it can explode in other more conventional ways from an enormous release of explosive gas, such as hydrogen that is produced during a meltdown. Engineers may calculate, correctly, that the chance of an accident is one in ten thousand years, but that is if all their assumptions of how the nuclear plant would operate are in place. When safety features are deliberately shut down (by engineers!) or when unpredicted damage to metal from exposure to radiation occurs, such calculations lose their meaning as the history of nuclear accidents has revealed. I have used my own rule of reason in assessing claims of safety — the more complex the system, the more difficult it is to account for all the ways that things can go wrong. No one designing the *Titanic* thought that the ship would graze an iceberg and be sliced open along its length, flooding the majority of its watertight compartments. None of the Soviet engineers designing the Chernobyl nuclear plants thought that their own engineers would turn off the safety devices while running an experiment on the graphite segments that were used to absorb neutrons. Similarly, none of the engineers and science consultants who helped to establish the Fukushima nuclear plants anticipated a tsunami that would breach the safety walls protecting them.

# The Limits of Our Assumptions About How Science Should Work

One good example of incomplete information will illustrate how the same data can be interpreted differently by experts. There has been an ongoing debate ever since Robert Thomas Malthus came up with what is called the population problem. Malthus, a critic of the idea of progress offered by the Marquis of Condorcet, argued that reason cannot lead to the goodness of society and humanity because biology would not permit it. He claimed that the means of feeding and nurturing humanity were generally linear in their rate of growth. Each person would need a certain amount of land to generate enough goods to keep that person alive. The size of that plot of land was assumed to be constant and the amount of arable land in the world is certainly measurable. But Malthus argued that most people in a reproductive lifetime can produce eight or more children and even if infant mortality killed half among the rich and poor alike, that would still double the population approximately every 20 years (or each generation). Population growth, he argued, was governed by an exponential rate of growth that would outstrip any linear effort at increased production of the goods for human survival.

In his own time, Malthus's most bitter critic was William Godwin, the father of Mary Wollstonecraft Shelley who authored *Frankenstein*. Godwin argued that Malthus erred in citing a doubling of population in North America but discounted the role of immigration in swelling the size of the population. He also argued that while some families had eight or more children, many more did not because of the death of the mother during or shortly after childbirth as well as a variety of complications that reduced the rate of birth after the first few children were born. He also argued that while the land may be fixed and the amount of land assigned could only be increased linearly, what the land produced could be increased much more effectively through improved agricultural techniques. Malthus and Godwin were both correct and wrong. Malthus predicted that England was incapable of doubling its population without massive starvation and death, especially of the poor. England has more than quadrupled its population since Malthus wrote his gloomy forecast, but he was correct that the North American solution, a Westward movement to reclaim land and dispossess the Native Americans, would lead to the virtual extinction of Native Americans. Godwin was correct in arguing that human ingenuity (using Condorcet's faith in reason and progress) would provide the technologies needed to improve the abundance of foods and necessities to feed an occasional larger population locally, but he was wrong in assuming that the world's population would remain fairly constant over centuries to come because of the factors he cited about how the surviving children who reach reproductive maturity could be maintained at replacement level.

The argument still runs today. Pessimistic scientists like Paul Ehrlich and Jonas Salk saw the population threat as approaching a crisis level and they expected mass starvation and shortages of non-recyclable resources. They advocated for birth control over all other remedies as cheaper, faster, and more accessible. Optimistic scientists like Amartya Sen pointed out that virtually all famines in the twentieth century have not arisen from droughts, overpopulation, or natural disasters but from political decisions that interfered with commerce, import of foods, or distribution of ample foods from one region of a country to another, or to ideological conflict where starving populations were subdued by withholding food.

While many scientists may agree that in principle an increasing population cannot be fed forever, they disagree that this will happen. They claim that literacy rates around the world and communication by radio and TV is so widespread that family planning is inevitable in most countries, with or without American efforts to export birth control. They also claim that family size, in the long run, falls when opportunities arise for surviving children. When children are almost certain to live to old age, parents feel less desire to have lots of children just to ensure that a few are left to support them in old age. These critics of population policy inherit the legacy of Condorcet.

## Science is Limited in What it Can Predict

Note how difficult it is to give precise answers to the issues raised by population crisis. How well can we predict the adoption of family planning in other cultures? How well can we predict the emergence of technologies that will increase food production for billions more people, or the greenhouse effects of people burning fuels to cook, heat their homes, or clear land for farming, or the rate at which other countries will adopt social services for the elderly, sick, and unemployed? One does not have to be a scientist to be a pessimist or an optimist. The public is as likely to be divided on population problems and policies as are scientists.

The social sciences include a range of disciplines, including sociology, political science, history, economics, anthropology, and psychology. Most of them lay claim to calling themselves social sciences rather than social studies because the methods they use are similar to those of the more traditional sciences, such as physics, chemistry, and biology. They use reason to gather and interpret data and mathematics to test the significance of their findings. Experimental psychology is extremely cautious in its experimental design and most scientists have no difficulty accepting the work as identical in kind as that done in other sciences. Physical anthropology, archaeology, and paleontology are essentially like the mainline sciences and use the same values. The other social sciences are less likely to use experimentation to test theories because it is difficult to do experiments with entire cultures or to find suitable controls for such experiments. We cannot imagine historians wanting to do experiments to determine outcomes of the future! I am very sympathetic to my colleagues who call themselves social scientists as they share the same considerations for the authenticity of evidence, reliability of memory, accuracy of eyewitness accounts, and similar concerns that make objective interpretations of cultures, history, and political systems difficult if not impossible. I only become troubled when they reject attempts at objectivity and label it political smoke screening to disguise their ideological bias, or when they claim that political clout rather than examination of evidence determines whose ideas are perceived to be correct and whose are rejected as heretical and not worthy of attention. It is not that such attempts at bullying people into belief do not exist. We saw this happen with Nazi racist science and Soviet Lysenkoism, but such outright pressure is far more difficult to achieve in democracies where scholars in almost all fields like to disagree. I hope that most of those in the social sciences do not share this view of a constructed reality through some consensus of the powerful. Such claims render it difficult to believe or even want to read publications by those who make them. In my mind, it would be like reading poorly written fiction (academic style on both sides of this debate is very turgid and awkward). Traditional scientists, of course, bitterly reject such an interpretation of science.

## Is it Fair to Describe Social Science as Dominated by the "Academic Left"?

Although Paul Gross and Norman Levitt discuss this issue in their powerful critique, *Higher Superstition*, their approach is a polemical, sock in the nose, gloves off approach to hitting back at their opposition — the world of the constructivist school that they claim includes ideological feminists, postmodernist literary critics, and an assortment of historians, philosophers, and sociologists of science. I find their term "academic left" unfortunate because it implies that this is primarily a Marxist-inspired group. While some of my colleagues who share these views are certainly products of the radicalization of student bodies in the 1960s, many more are not. They genuinely believe there is no reality out there that can be described. Whatever is out there may be real, but no scientist can successfully use words or the tools of science to describe that reality. All reality, in their mind, is a construction and it is temporary, a paradigm waiting to be replaced by another way of wording it by some future generation of scientists. I think of these people primarily as skeptics and philosophically closer to nihilists or solipsists, the former believing in nothing and the latter at least acknowledging their own existence or self-awareness. I wish that Gross and Levitt had just called these scholars members of the constructivist school instead. Perhaps they felt that exposing what they thought was these scholars' real intent — a radical reshaping of society — was more effective than staging yet another academic debate that the public would not be interested in following.

In my youth I would have found it strange to hear the phrase, *social sciences*. My high school teachers taught *social studies*. History was considered one of the humanities because it dealt with the issues that literature took to heart. Economics had no pretensions then of being a science; at best it was a heavily mathematical analysis of money, resources, and manufactured goods. Even my math teachers did not consider themselves scientists. They talked of mathematics as the "language of science." Psychology was considered a science because it has a long history of using experimentation.

If I were to offer advice to the public on interpreting science, I would urge some sort of reserved judgment. Not all science is good, which is true in all fields from physics and biology to economics and sociology. Good science is subject to testing and generates predictions that can be verified. Good science drives out bad science. The existence of some bad science (like the negative eugenics in the first half of this century) does not make all science in that field politicized or bad. The working out of the structure of DNA is a triumph of human intellect and reflected virtually no socialist or capitalist ideology. The personalities of Watson, Crick, Wilkins, and Franklin may have differed across a political spectrum, but their differences as scientists were altogether different. They reflected openness to unconventional approaches, habits of discourse, differences in imagination, and intensity of effort. Pessimists and optimists, enthusiasts and wet blankets, the morbidly shy and the horn-tooting egotists, and those with flamboyant personalities and phlegmatic personalities exist in all ideologies, cultures, and fields.

# 16 Science and the Supernatural

Science likes to distinguish the world it works in, the natural world, from the world it cannot work in, the supernatural. The natural world is also the material world — composed of atoms, atomic particles, radiations, and laws that describe how all of these are organized, work, and produce the phenomena we call the universe from quarks to quasars, from the incredibly small to the incredibly large beyond our imaginations. Outside this immense realm in which much remains to be learned, is a world closed to scientists. It is the realm of spirits, souls, and gods; a world that requires faith for its acceptance because, however much we might infer about the existence of a supernatural process or thing, we cannot demonstrate it by the tools of science. The supernatural is, by definition, beyond the natural or material world accessible to science.

## Claims of Supernatural Phenomena Lack Scientific Support

In contrast to science, which is accessible to everyone who wishes to learn how to engage in it, use its tools, understand its careful procedures, and achieve a level of competence to do some branch of science, comparable access is not possible for those in contact with the supernatural world. Not everyone has extrasensory perception; not everyone can make dice and other objects move at will, bend spoons by looking at them or gently stroking them, or will stopped watches to begin ticking again; not everyone receives dreams, visions, or second sight of events occurring at a distance or some future date; and not everyone can read someone else's mind or experience a déjà vu.

For the person who has these experiences or witnesses them in others, these are as real as the moons Galileo saw around Jupiter. For many scientists, there is less conviction that they are real phenomena and for almost all professional magicians there is no conviction whatsoever that these claims of supernatural phenomena are real. Quite the contrary is true. Magicians like James (The Amazing) Randi have made a profession dogging the heels of faith healers, spoon benders, mind readers, readers who see through blindfolds, and a variety of others who claim special supernatural talents. In every case, he has duplicated the tricks used by alleged psychics and by those who are self-deceived. He has a standing offer, now some $100,000 or more, for anyone who can come to him with a supernatural talent that he could not duplicate under controlled conditions set up by him and his fellow magicians. No one has yet claimed that money although many have tried.

## At Best Science and Religion Try to Practice Tolerance

While scientists are only too willing to expose fraudulent practices among these claimants of supernatural gifts, scientists usually feel more ambivalent about the supernatural in the form of religion. Here, respect for religious belief is important. Religion has a long history stretching across millennia, although few of the religions today exist in the forms they had back in the past. Most are just demoted to the status of myths, including the polytheistic religions of Pharaonic Egypt, Golden Age Greece, and the millennium or more of the Roman Empire. Religion for most of humanity provides a sense of purpose, moral code, belief in an afterlife, belief in a Creator of the universe (usually God or some polytheistic member of a divine family), and community or fellowship that gives people the satisfaction of belonging to a common faith. There are immense variations in theology, rites, rituals, obligations, statements of faith, and other traditions that adhere to religions. Yet among hundreds, if not thousands of different religions in the world, any observant person will feel as assured of the goodness of this one particular faith as citizens generally feel about their communities or nation. There are a few religions that do not worship a god, do not demand a statement of faith or belief, and actually reject the supernatural or leave that question open to personal conscience. Such religions rarely create difficulties for scientists because they do not demand some supernatural components.

Scientists are largely hostile to claims of the supernatural although some very prominent scientists believe this to be an open issue and do not wish to reject possible psychological findings that a study of parapsychology might reveal. They are much more divided about the claims of the supernatural in religions. Many are atheists, many are agnostics, and many are dualists who live happily with their science and their religious beliefs and activities. I have met prominent scientists (members of the National Academy of Sciences, for example) who are religious. They serve as deacons of their churches and consider themselves observant as Catholics, Jews, or Protestants. I have no doubt that this is true among scientists in other cultures and there are likely many Hindu, Buddhist, Moslem, and individuals of other major religious beliefs who are accepted by scientists in these cultures.

Sometimes religion and science make it difficult even for a dualist to accept both ways of interpreting the universe. It would be almost impossible, for example, for a biologist to accept natural selection and life originating from non-living matter some billion or more years ago while at the same time believing in the fundamentalist Protestant claim that the universe is about 10,000 years old and that all of life as we know it today was put on earth during the first six days of the existence of the universe. It would also be difficult for geologists and astronomers to accept such fundamentalist beliefs. I do not doubt that there are such individuals who try to straddle both beliefs and who rationalize away their science (or their religion) in some way to find harmony in both their professional lives as scientists and their private lives with their families.

One reason why a scientist can be quite proficient working in the material universe and at the same time be devoutly religious is that the two are often as immiscible as oil and water. We are good at balancing contradictions in our lives. The world is necessarily more complex than our efforts to simplify or explain it suggest. We use one exception after another to justify killing but continue to condemn murder. We believe in peace but come to the defense of our country when we engage in war.

We divide our loyalties erratically among contradictory demands of family, work, profession, community, religion, and nation.

Despite that ease with which we can live a life of contradictions, scientists find it difficult to accept claims of the supernatural, mysticism, holism, and other unprovable features of the physical and living universe. This is a long tradition in science. Scientists quickly learn that they can find, prove, disprove, and analyze things using the rational methods of science. They learn quickly that this other supernatural world does not yield to the same bag of intellectual tools. If the supernatural indeed exists, it then has to be held in some kind of faith that ultimately its material basis will be worked out and the paranormal phenomena will become normal phenomena. That is faith, which makes a scientist uncomfortable because it requires the scientist to mix religion and science as if they are compatible approaches to understanding. This would require withholding the tough standards that are used in the material world out of sensitivity for the feelings of others when it comes to matters of the supernatural.

## Few Today Would Accept a Platonist's View of Reality

I am much instructed by the way reality was regarded in the past. For many Platonists, the real was not what we saw, felt, or described before us. It was rather the ideal form of what we saw or experienced, whether it was a chair, a pencil, or a circle. All variants of these objects were imperfect representations of the ideal and hence not real but corruptions of the ideal or temporary modifications that had no persistence. Ideal chairs, pencils, and circles were immutable and universal and would never change. Hence, they were the real while the world of things we perceived were the unreal material world. Most scientists, and most of humanity today, would utterly reject such a philosophy. We are too much of the world and too secular in our values. We like the material world of ephemeral things and we do not believe that we are buying illusions but rather material goods that give us comfort, pleasure, and necessities. I remember many years ago when a high school friend was doing his doctoral work in philosophy at Columbia University. He went with some faculty and fellow graduate students to a local restaurant and they thought they would tease the waiter. They asked for the soup of the day and after being told it was onion, pea, or lemon chicken, one of them said, "We don't want a modified soup; we want plain soup." The waiter hardly batted an eyelid and replied, "I'm sorry, Sir, we're all out of plain soup. We only have onion, pea, or lemon chicken." You don't have to be a professional philosopher to get out of a logical trap!

## Creationists Claim to Follow the Procedures of Science but Courts Have Held Otherwise

A few years ago, I debated Duane Gish of the Creation Research Society. He had actually wanted to debate a colleague, Douglas Futuyma, who had written several books on evolution including one, *Science on Trial*, that attacked the claims and tactics of this group. The creationists of this particular group claimed that they were all scientists and that their argument was not one of religion against science but of science (their approach) against science (our more rigid approach). I chose a different way to approach the debate because I learned that Gish had debated more than three hundred times, and this was my first opportunity to do so. If Gish had the advantage of familiarity with his arguments, I had the advantage of an auditorium where I was comfortable talking to 500 or more students. While quite a few Stony Brook University students did show up, there were busloads of students from churches or other campuses that were very committed to a fundamentalist world view. My most novel approach was to examine Gish's claims on his own terms. If creation science was a science and not a religion, how did traditional science, traditional religion, and Gish's creation science agree or disagree on such aspects as source of evidence, authority, reliability of evidence, testing of evidence, response to contradiction, places of publication, citations of published work, and similar attributes of scholarly work? It was very clear that most of the methods used by Gish and his supporters were comparable in style and procedure to those in traditional religions. They were quite different from the criteria used by most scientists.

Science would have a difficult time if it had to follow the constraints of different religions. Thus, if miracles are historical events and the work of God, they cannot be duplicated by scientists and many cannot be explained away by scientific means. The Bible abounds with such miracles. No geneticist, not even a Lysenkoist, would believe that Jacob's painted rods would create some genetic impression on Laban's mating sheep and result in variegated coat color. It is equally difficult for most physicians and microbiologists to believe that disease is a consequence of sin or lack of faith and the cure of illness is neither science nor a miracle, but a faith in Jesus. Many Christians reject germ theory or traditional medicine and believe that faith will cure their cancers, paralysis, diabetes, failing kidneys, blindness, or other ills. It is this faith that makes the desperately ill become gullible victims of faith healers who are either self-deceived or out and out frauds. Any psychic or faith healer who claims that true miracles of recovery have occurred from fervent faith would find it difficult to pass scrutiny from Randi and his team of investigators.

## Galileo's Trial for Heresy Soured Scientists on the Authority of Religion

Galileo's plight has been an object lesson of the conflict between science and the authority of religious faith. Today the Catholic Church is much more cautious in taking sides on scientific findings. It acknowledges, for example, the legitimacy of natural selection to explain the evolution of life as long as it does not discuss the evolution of the soul, which, of course, it does not. Nevertheless, Catholic biologists would find themselves in difficulty with their church if they did research with preimplantation embryos or the development of new techniques for reproductive medicine. Such violations of Aristotelian principles of natural law would be considered inappropriate, especially if this were done in the laboratory of a church-affiliated school.

Orthodox Jews as scientists would find similar restrictions from working on the Sabbath to confronting Talmudic perceptions or interpretations of individuals with birth defects, especially those associated with genitalia. Whether a hermaphrodite or pseudohermaphrodite (now called intersex individuals) repaired by surgery and maintained by hormones would be given full status for marriage and participation in temple rituals is doubtful and would involve considerable debate in Jewish courts, which would most likely rely on the authority of Talmudic commentators than the authority of contemporary physicians or embryologists.

## Science Weakens Belief in the Supernatural

The history of biology illustrates well the way science erodes supernatural beliefs in the life sciences. Until the mid-nineteenth century, there were few attempts at experimental biology. There was a lot of good descriptive biology, and there were also hints that the body could be seen as a complex machine. It must have been a sobering experience for those who first read, in 1628, William Harvey's studies of the circulation of blood, his thorough demonstration of the heart as a pump, and the volume of blood it squirts into the aorta each day. It would be more than two centuries before life in general got shifted out of natural theology and fully secularized by biologists like Darwin. While a God-permeated life became less fashionable in the nineteenth century, there was widespread belief that the imprint of the creation of life made it inaccessible to scientific study. By this time, most scientists embraced reductionism, an approach that assumes that a complex system can be analyzed piece by piece and that the entirety can be understood from the components, including the processes that make the system work. Reductionism in physics, geology, astronomy, chemistry, and biology produced immense insights into the composition of the universe and revealed many of its processes. Reductionism is also the universal approach of engineers. They are as fully aware as they can be of a complex structure and how to make it functional and safe whether it is a bridge, a steam ship, a dam, or a factory. In the nineteenth century, the success and pervasiveness of reductionism in the industrial revolution and in science led to a revolt by those in literature. We see it when Walt Whitman walks out of the lecture by "the learned astronomer," in the lengthy polemic against "Aries Tottle," in Edgar Allen Poe's prose poem, *Eureka*. We feel the anger in Charles Dickens' treatment of Coketown and the utilitarian and emotionally dead world in *Hard Times*. I feel myself slapped in the face as I read the description of the horse by Gradgrind's teacher's pet ("a gramniferous quadruped") as well as Sissy Jupe's portrayal of horses as companions who extend human abilities and give in kind the love they receive from their trainers at the circus.                    .

## The Life Sciences are Often Targeted for Injections of the Supernatural

There was also a resistance by many biologists. They may have rejected Paley's approach of seeing trout leap out of the water to express their pleasure at the happiness that God gave them for their being alive, but they felt strongly that reductionism had its limits when dealing with life. One belief was that chemistry did not apply to living material and reductionism would not yield how the life stuff (vaguely called protoplasm by Huxley) worked. A second belief was that any attempt at reductionism at the cellular level would fail because the tools of reductionism would destroy the living state. The underlying assumption of this belief is holism. For living things, the whole is allegedly greater than the sum of its parts. A third belief invoked some supernatural component sometimes called "élan vital," or mneme, or enteleche. Whatever the name of this inaccessible supernatural force, it constituted what its critics called vitalism, which assumes that the breath of God is in any living material and that science cannot use reductionism to locate it because it does not belong to the realm of science.

All of these beliefs crumbled as reductionism, to the disgust of its critics, triumphed. Organic chemistry developed out of the study of molecules found in living things and biological products like urea were soon synthesized by chemists. Out of organic chemistry came biochemistry and the study of enzymes, metabolism, the composition of proteins, carbohydrates, and nucleic acids. Cell biologists in the twentieth century learned to isolate organelles of cells and work out their functions. The cell's harvest of information yielded readily — the cell membrane regulated how molecules used energy to move molecules in and out of the cell; mitochondria were shown to take the oxygen we breathe and produce energy from oxidized foods and trap that energy in chemically useful form; lysosomes were shown to be small bags of enzymes that cleared out wastes and recycled bits of the cells; the endoplasmic reticulum turned out to be the place where protein synthesis occurred. This is bread and butter science. Reductionism is the way science works at its best and scientists do not need to apologize for using it, however uncomfortable that word sounds to those who want a world that contains mysteries. There are enough complex systems that will challenge reductionists for generations to come. No one, for instance, has devised an effective computer model for figuring out how proteins fold into their three-dimensional forms. Many systems, like weather patterns, cannot be predicted more than a few days in advance. We know virtually nothing about the molecular basis of learning, memory, thought association, and creativity, let alone self-awareness.

## Reductionism Usually Prevails Over Vitalist Interpretations of Life

The history of those who reject reductionism and embrace some form of vitalism, holism, or mysticism residing in the deep recesses of life has been one of retreat. Each falling back leads to a new standard of where the mystery resides. The conquered outposts of the body, organs, tissues, cells, organelles, macromolecules, and similar reductionist assaults that yielded an understanding of basic life processes are ignored as if they never were once part of the periphery that separated knowledge from mystery. I have long puzzled over this need to assign mystery to complexity and to somehow protect ignorance from yielding to scholarly and scientific study. I find quite the opposite emotional feeling as knowledge reveals how things work and how things are related. I delight in being a community of many trillion cells and thrill at seeing that my DNA is related to apes, mice, flies, worms, mushrooms, and yeast cells. It extends my sense of kinship and my appreciation of our common life in its long journey over a billion or more years on earth. I am excited to read that the heavier elements of my body are descended from a dying star whose remnants formed the earth and its sibling planets. Learning that I am connected to the universe in that way, as Carl Sagan so eloquently presented in *The Cosmic Connection*, greatly enlarged my sense of what it means to be alive and be both a part of and apart from the universe. It is not mystery that resonates in my being but new knowledge and ways to perceive myself. I too can exult like Walt Whitman and proclaim that I contain multitudes!

Science does not deaden the soul any more than living in a secular world might. Most of humanity enjoys worldly things and rejects asceticism. We can appreciate our lives and find a place in the universe without withdrawing from the world, flagellating ourselves, mortifying our flesh, and accusing ourselves that we know too much. To live in the world and not only enjoy but also understand it does not necessarily make us depraved or immoral. Quite the contrary, we learn that the more of the universe we understand, the more enriched our lives become, and the more we want to protect this world from abuse and pass on this blessing of knowledge for future generations to enjoy.

While scientists frequently appreciate the benefits of new knowledge, many non-scientists frequently look at advances of knowledge with fear or disapproval. All knowledge can be abused, regardless whether it comes from science, philosophy, or theology. Wars have been waged in God's name and many have been slaughtered for their refusal to convert to a conqueror's religious beliefs. Ideologies that seem universal and true to their political supporters have been just as mischievous in demanding conversion or at least silent acceptance of communism, fascism, or other totalitarian systems. Science is not immune from abuse. The deep resentment of science I encountered among students who feared or avoided it was often due to its lack of moral concern to its uses. When scientists look the other way as their instruments are used to subdue people in war, they share the same habit of mind as a German populace looking the other way as their Jewish neighbors gradually disappear from their midst. Scientists are as vulnerable as any citizen to prioritizing patriotism over conscience, or job security and group loyalty over the safety of laborers or a local community at risk. They too can look the other way as beautiful neighborhoods become bleak and soulless Coketowns, begrimed by industrial wastes. The public has a right to protest the indifference of those who use science to create technologies that deaden our lives and alienate ourselves from what is beautiful and decent.

This indifference is not so much a uniquely scientific blight as it is a fault of those who are apathetic, have no sense of wonder, or do not see themselves as connected to others and the world in which they live. The indifferent may have little appreciation of science and their pleasure may be in power or the accumulation of wealth. It is also a fault as old as the recorded history of humanity.

## Science Changes the Worldviews of Civilization Through its Rejection of the Supernatural

While many of those who dislike or fear science base their attitude on the failings of scientists to use their talents and discoveries for good, there are others who dislike science because they see it as diminishing the realm of the supernatural. There are good reasons why this fear is justified. In the fourteenth century, there would have been very little of the universe that was secularized. Belief in God was universal or nearly so, and all of knowledge was essentially bound to the Church. A major belief was that life on earth was inconsequential and merely a test to determine whether one would live in eternity in Heaven or Hell. In a world where ignorance prevailed, certainty and knowledge seemed to reside in both the church and the mysteries that were denied us while alive on earth. Many people today still long for the mysterious and find great comfort and emotional fulfillment in seeing it as a sign that the holy exists. Historical examples abound of the presence of the divine in the mysterious, from God's presence in a burning bush to his presence as a voice in moments of revelation. For those who appreciate this mystery in things, the removal of mystery from rainbows, the movement of planets, the explanation of comets, the source of volcanic eruptions, the origin of hurricanes, the atomic composition of all material objects, and the explanation of heredity through the gene as nucleic acid shakes the foundations of their understanding of reality. The loss of mystery renders their universe void of encounters with the divine, reducing human life, dignity, and culture to a heap of material rubble barren of celebration and wonder. This, of course, is false. The problem is not so much science as a culprit but rather the more enduring problem of human temperament. Every generation contains unhappy people who fix blame for their discontents on others and not themselves. These unhappy individuals are vulnerable and threatened because they have inadequacies or do not know how to cope with change. I am not talking about the oppressed, the impoverished, or others who indeed have reason to complain, but rather those whose basic necessities are satisfied but whose inner needs are wanting. Such individuals cut across all disciplines and are as likely to be found among scientists as among those in the arts and humanities.

The supernatural has its appeal because for most of humanity, the world and even the universe of things and explanations are not enough. Many people, including some scientists, have a deep need for religious experience. They enjoy a belief in a life after death and find the idea of dying and having no personal soul and no God to encounter a terrifying if not bankrupt thought. Religion is a necessity for them and science will always be a threat or a limited part of their universe. The business of science is neither to make the world atheistic nor to placate the human need for religious belief by withdrawing from studies of astronomy, geology, or life that offend those whose religion demands a non-scientific worldview. It is not that this demand isn't desired. I once gave a talk at a church-affiliated liberal arts school in the Midwest, and one student berated me after I had finished: "Why don't you scientists work on things that help us live better lives, cure our diseases, and things like that? Why do you try to compete with religion and talk about doctrines like evolution that aren't science but just theory?" Such a defensive shield against knowledge is difficult to penetrate. There are many people like this student who fear that science destroys religious belief (and indeed they are right) and thus they resist science as the enemy of their beliefs. My one comfort when I reflect on my inability to reply effectively to such criticism is that the shield encompassed a far larger segment of the world previously and that scientists back then were not only perceived as heretics but also at risk of arrest, inquisition, or punitive condemnation.

# 17 The Pleasure of Science

I love being a scientist. It has given me immense pleasure to read, think, do, teach, and communicate science. In this respect I am not much different than any person fortunate enough to love doing what they do. Many of my colleagues in science feel the same way and I recall how often we would remark, "I can't believe I'm being paid to do something I love." Science today is highly specialized and there are few scientists at home in different fields of science. In general, scientists can only read superficially about fields they have not been immersed in. The science that one learns as a physicist usually does not carry over to molecular biology or population genetics. It is often with difficulty that I go through an issue of *Science* or *Nature* and try to understand more than just the abstract of articles from fields I am unfamiliar with. Most scientists find themselves cut off from other sciences for that reason. Things were quite different 100 or more years ago when a scientist could read articles from almost any discipline and still glean a good deal from them. The age of Humboldt and Darwin is no more as universal learning has given way to specialized learning. To maintain some connection to the rest of the knowledge universe, scientists depend on review articles, commentaries about science and the news, and filtered versions made by gifted popularizers of science.

## Scientists Must Read a Lot to Keep Up With Their Fields

That's not bad. Scientists still learn a lot about happenings in fields they cannot read with the accuracy and critical judgment they are capable of in their own area. I try to read about what is going on in many different fields including geology, astronomy, physics, and chemistry. I like the way my own field, genetics, gives me an entry into reading the medical, biochemical, cell biology, and evolutionary biology literatures. Scientists are very curious about the universe and this makes them read widely. But doing science requires a lot of skills that reading alone cannot provide. The scientist goes through an apprenticeship, is weaned on simple uses of machines and other scientific tools, and learns to apply them to simple problems. Every research graduate student has seen hours of work go to waste because the techniques and preparation were faulty. Learning the pitfalls of techniques and how to master them is part of doing science. This is why it is so difficult for the lay person to appreciate how science is done. In many ways, it is like learning a musical instrument and trying to perform it well. We expect such practice and mastery in the arts, but we often fail to appreciate how much repetition and learning goes into the techniques to perform science. I recall my own frustrations with looking at tissues and cells when I tried to spot the differences that the text said were there. I saw that as I became a teaching assistant and witnessed the variation from specimen to specimen as I walked around the laboratory classroom. It's not that I was obsessed with such detail. In fact, I used to arrange my walk past the tables so that I came out at the door of the laboratory preparation room and set up a chess board and made my move and then left out of the other door that got me back to the first table again. When I wasn't focusing on the common carotid artery, I was focusing on my next best move in the game I played with another teaching assistant who was on the opposite side of the room.

## Scientists Sometimes Play to Refocus Their Minds

The need to play and to do other things while heavily immersed in doing science is widespread. I do not know of a fruit fly laboratory where the graduate students have not engaged in rubber band wars at some time or other. Rubber bands are typically used to keep bundles of vials together, but very quickly I learned to fashion a rubber band gun by stretching a rubber band from the tip of my extended index finger around my thumb to the small finger on the ulnar side of my palm. The sport was to shoot house flies (not our precious fruit flies) that buzzed about on hot summer days, but sometimes the target was a fellow graduate student. To improve aim, we learned to stretch a rubber band on a twelve-inch ruler and this greatly extended our range and accuracy. I learned that those rubber band wars date back to Morgan's laboratory and offered relief from the tension of spending hours hunched over the eyepieces of a dissecting microscope counting flies.

Doing science is fun as well as hard work. You see results coming together whenever you have a good experimental design and know what to look for. I frequently thought of those flies as I counted, sorted, and recorded them before disposing them into a jar of alcohol and watching the etherized flies sink into oblivion. They were laboratory fruit flies whose ancestors were first tamed some forty or more years ago, many of them descendants of distant ancestors in Schermerhorn Hall where Morgan and his students worked. They had been mailed in vials or plastic tubes across the United States and around the world. Some 800 or so generations later, they were on my porcelain plate, separated into rows of males or females or rows of recombinant classes as I searched among them for signs of a crossover occurring within a nest of closely related genes. At two or three in the morning, I would feel as if the entire universe was in this one sanctuary that was shielded from the world, and everything that I cared about was focused on that plate no bigger than an index card. Whenever a rare event occurred, which was about once among several thousand flies, I felt a rush of excitement. The fly that I had sought (or that was a total surprise) emerged, and I carefully entered the information about its genetic markers, swept it into a vial with an appropriate fly to mate with, and maintained that fly as a stock. Each of these rare events caused me a great amount of elation.

I was not alone in this experience. All of my fellow graduate students felt the same way when they were immersed in work. We loved being in the laboratory and it was with difficulty that we interrupted our work to go to seminars, attend classes, or teach students. The laboratory was our domain; it was where we were seeing portions of the universe no one had glimpsed before, our holy temple where science was done.

## Scientists Take Pleasure in Discovery Because it is Rare

Nothing matches the thrill of discovery. When something new arises because you predicted it from prior work or because it is a totally unexpected surprise, you feel a long-lasting emotional euphoria. No stimulant has the same enduring feeling. If these are self-induced endorphins that tickle the neurons, they are both abundant and intensely pleasurable. It is more satisfying than doing an experiment and seeing it through to its outcome. It is the knowledge that you are alone, among all seven billion people or among all the people who have ever lived, in seeing something that no one else has ever seen before. No matter how minor the contribution to knowledge may be, it is unique and for at least a brief moment, you know that a bit of the universe has been revealed.

The urge to share that discovery is overpowering, which is why scientists write their findings in the form of scientific articles. They are published in journals that are relatively obscure to the general public. As we had noted earlier, authors receive no payment for the article and in fact may have to pay "page charges" if they have a grant from a federal or private source. The author may also send a few copies of the article (called reprints) to fellow scientists who are working in similar research. This practice was much more common in the days when duplicating technology was limited to mimeograph machines and actual photographs of the pages of the article. Today, printing facilities allow for rapid duplication and most scientists do not bother writing to colleagues for a copy of their work; they just make a copy for themselves using a Xerox machine in their department office. They know that others read their articles when they see their work cited in the references of other scientists. There are also citation indexes today that inform both the author and anyone who wants to know how often an article is cited and where. A few superstar papers have citations in the thousands while most others are lucky to have a dozen or two. It does not matter to the scientist who publishes. The work may be esoteric, but the fun of having found something new and worthy of publication is itself a heady experience.

# Scientific Meetings Lead to the Stimulus of New Ideas and Findings

If scientists do not write up their work for publication in professional journals, they sometimes present their work at international meetings and enjoy the pleasure of discussing their work with other colleagues. A few scientists are painfully shy, have stage fright, or are inarticulate, so such an experience would be an ordeal for them. These scientists often choose to do their research in research institutes rather than universities so that they do not have to teach and agonize over facing an audience. Most scientists are not afraid to talk about their work and they are used to lecturing students. Some are spectacular at presenting and have a gift for popularizing their own ideas. Muller did both. He could sometimes overwhelm an audience with a technical paper and would seem to go on interminably. At other times, he could present his ideas with such clarity that the audience was spellbound. When I went for my job interview at Queen's University in 1958, Dean Rollo Earl told me of the excitement he felt when he was a young scholar in Toronto and heard Muller present his 1921 paper on *Variation due to change in the individual gene*. He said that he could still feel the thunderbolt of surprise when Muller said, "Let us hope so," after urging geneticists to study the gene by becoming physicists and chemists as well as biologists.

I have met other scientists who like doing experiments but hate to write up their results. I know of one student of Muller's whose magnificent dissertation on spontaneous mutation frequency has never been published. One of my own students never published his superb PhD dissertation work on the cytogenetics of the dumpy locus. A brilliant colleague at UCLA abandoned his university career for a full-time research position in Switzerland where he has flourished without publishing much. He feeds ideas to others who write articles and make him a co-author. He never felt threatened by this inadequacy and was happy being a stimulant for his colleagues and students. He shared a similar personality to Max Delbrück, who did publish a few key papers and founded a field but limited his attempts at publication to articles that he felt had something truly new for others to read. Extending what is already known was not his interest.

## Each Field of Science has its Own Writing Style

Scientific writing is more compressed and less literary than other writing. There are many reasons for this. Journals want to economize words, especially when articles are costly to produce. The scientist also knows that virtually no one other than scientists in the field will be reading these technical articles. There is no reason to provide historical background or detailed descriptions for terms, processes, and hypotheses that are well known to those readers.

Publishing adds to the prestige of the scientist because all professional journals require peer review of the articles submitted. Usually two persons review the article, and the editor chooses reviewers who are knowledgeable about the topic. This process can occasionally lead to disputes if the reviewer is from a rival camp and the author feels that the article is treated unjustly. If the author complains, a third reviewer may be chosen. While it may not be a perfect system, it is essentially fair and significantly better than simply publishing any articles submitted, which would result in a lot of sloppy work published and be enormously costly. It is also fairer than giving that task to one editor or a small panel of experts. They might have too limited a bias. Most scientists are elated when an article is accepted for publication as their contribution to knowledge will now become known. When something is published, it will be available in libraries for decades or even centuries to come. Even if the articles are rarely cited some five or ten years later, there are always historians who go back to review the field or trace the history of an idea or discovery. It is always fun to hear of how scientists responded to their first publication. Copies have been sent to puzzled but proud parents, grandparents, relatives, and close friends!

## Scientists are Most Often Rewarded by the Satisfaction of the Work They do

Fame is the good fortune of a few scientists. Most are not famous and have little chance to ever become famous. Of those who became famous, most did not actually seek it. What motivates most scientists is the excitement of working in a laboratory, in the field, or at a desk. It is where the work gets done and discovery happens. It may come in the form of a new theory, finding, process, or technique, or some other remarkable new thing that changes the way people do science. Students are seldom motivated to enter science to become famous, get rich, or win a Nobel Prize. This motivation is also true for admission to medical school. Medicine is an applied science and most physicians want to treat patients. Although critics often see medical students as seeking power, money, and prestige, most of the medical students that one gets to know well are decent human beings, exactly the kind of people you would want to care for your own or your family members' health. Scientists, too, fall into a simpler class of people who are motivated by the thrill of learning and contributing to knowledge through research findings. Science always has discovery as one of its aims. Those who apply it, like engineers and physicians, are constantly making discoveries that they eagerly share.

There will be some scientists who are competitive, seek recognition, thirst for power, and may spend disproportionate amounts of their time in the politics of science. This inclination exists in almost all human professions. It does not matter whether such people become famous and make major contributions or not. Most scientists are not initially motivated by these other interests. I have only encountered a few who act in such a way or admit to having such a personality. There are pressures on scientists that may make them alter their pace of doing science and prioritize efforts to advance their careers. All university scientists know that they must publish their work in order to be promoted. Without tenure, there is a great deal of unhappiness for the academic scientist who does not have another shot at a tenure track job. Most scientists who like the academic life end up in community or four-year colleges where very little research is done.

## Fame is Rare and Usually Does Not Endure

Of the tens of thousands of scientists over the past two centuries, there are fewer than 100 who are household names and whose work, in their own lifetime, has been received with enthusiasm, prestigious awards, national honors, and public accolades. We think of Pasteur and Koch, of Lister, of Roentgen, of Einstein, and of Darwin. Once we exhaust the familiar names of Planck, Bohr, Rutherford, Curie, and a few dozen more, we realize how fleeting fame is and how short a time it endures save for a rare few. Fame rarely dominates the ambition of scientists and most know that fame will elude them. What scientists crave most of all is peer acceptance. When the young scientist is greeted by name by a Nobelist or a member of the National Academy of Sciences or the Royal Society, this creates a feeling of having arrived, of being recognized as a contributing member of a field. Scientists often reject public recognition as shallow, adulatory in a pathological sense, and dangerous. It is not the public that scientists go to for attention and approval, it is to their fellow scientists. I remember going to an invited meeting at Lake Arrowhead and presenting my work as a young scientist at UCLA. Scientists at Caltech and Stanford called me by my first name and talked to me as an equal. It felt unreal, as if this could not be happening, but it was also a wonderful feeling knowing that I was a colleague whose ideas they valued.

## Scientists do not Always Appreciate the Need to Communicate Science to the Public

It is sad, however, that scientists who genuinely love science and want to communicate ideas to the public are sometimes vilified or shunned. Popularizing science is often seen as pandering to the masses, especially if the scientist writes a literary work for the public and it becomes a best seller. Carl Sagan's career was hurt by such public acclaim as many of his colleagues felt that it was inappropriate for him to go on television and appear on the Johnny Carson late night show. They saw his public television efforts in the *Cosmos* series as a form of grandstanding or puffery for public recognition, rather than the powerful educational program that inspired thousands of young people to consider science careers. It was ironically Sagan's popularization of science that helped some of his most vocal critics get the grants they needed for their research. Congress heeds the people and when there is a strong demand by the public to support science, science will be served. One way in which this undercurrent of disapproval damaged Sagan was his being passed over for membership in the National Academy of Sciences. He never received an elected membership allegedly on the grounds that his serious work as an astronomer did not live up to the standards of his peers. Only very late in his career was he given public service recognition by the National Academy of Sciences. Fortunately, Sagan was well recognized and appreciated by the students who came to Cornell to work with him as well as his colleagues. Many Americans are also grateful to his concerns over the potential disasters that await us in an atomic age, in particular his provocative analysis of "nuclear winter" — the synthetic atmospheric disaster that might occur in an all-out nuclear war, which would leave in its aftermath a world not too different from that which followed the meteoric impact some 60 million years ago that led to the extinction of the dinosaurs.

## Scientific Findings Often Evoke Wonder and Awe

Science also gives all of us pleasure. We delight in new knowledge, and I love to watch as children make their way through the exhibits of dinosaurs or the evolution of humanity in the American Museum of Natural History. They stare wide-eyed as they realize how the earth once looked, and the gifted artists who depict the dioramas and murals to illustrate these lost worlds give our imaginations a leap into the past and demonstrate how scientists have retrieved a part of our forgotten past, our legacy, and our extended family of long lost kin.

# 18 The Curiosity of Science

Scientists are nosy and have a curiosity about everything, like children. For most people, that curiosity about the universe gets blunted over time, and for a few, there is very little left of that wonderment about the natural world. Curiosity shifts to politics, sports, family, or the workplace. It may be satisfied by watching soap operas, reading mysteries and thrillers, or seeing films and plays. For the scientist today it is intensely focused on a segment of the universe. For about two years of my life, the most important thing in the world was the structure of a gene called dumpy on the left arm of the second chromosome at 13.0, or what is sometimes referred to as the prominent shoe buckle band of the giant salivary chromosome of *Drosophila melanogaster*. Pretty esoteric stuff, isn't it? But for me, it was a pleasure — a passion even — to induce dumpy mutations with x-rays, map them, and figure out the structure of this gene. It was new and different from all other attempts by my colleagues in other universities, so I was adding to their knowledge with my work.

## Why Scientists Should Focus on a Small Segment of the Universe

Scientists may restrict their focus to a very specialized bit of knowledge and magnify its importance in their mind, but this is not as silly as it sounds. Our sun is one out of a hundred billion stars in our galaxy, the Milky Way, which is one out of billions of galaxies in the universe. Why then do we hold the earth, one of just nine planets spinning around the sun, and one of middling size, so dear? We learn a lot about the universe no matter what tiny bit we focus on. Most scientists are specialists for this reason. They can learn a lot from paying attention to one small aspect of the universe, which is the ultimate triumph of reductionism. At the same time, although the scientist is hacking away at something seemingly minute, the information gleaned can be connected to a large body of knowledge and thus the findings may take on cosmic proportions. Look at DNA for instance. There are some three billion nucleotides stuffed into the nucleus of a sperm or egg. Organized as genes, they determine what the resultant individual will be — plant or animal, mammal or invertebrate, human or rat, male or female, light skinned or dark, hairy or sparsely haired, of normal abilities or impaired in some serious ways. The scientist who is looking at the nucleotide sequence and focused on just one of those genes may tell prospective parents whether the embryo in the mother's womb is normal or has cystic fibrosis. For the parents, this information is monumental. For that week or so while they wrestle with the information, this tiny segment of the universe would be the most important thing in their life. Suddenly, this initially insignificant bit of matter, this trivial fraction of one percent of their nucleotides, becomes a touchstone to their fundamental values and beliefs.

## Scientists Seek the Significance of the Work They Do

Science has such appeal to scientists because curiosity goes with imagination. Scientists not only want to know about something that is of interest to them, they want to milk its significance. Scientists can think ahead hundreds, thousands, or even millions of years, or travel back over similar stretches of time as they plug a bit of knowledge into some broader picture. They can also use their imaginations to relate their work to seemingly unrelated fields. There is some degree of Sherlock Holmes in every scientist as bits of information become clues. What was once a puzzle or unsolved mystery becomes clear. All scientific imagination is bridled. The scientist differs from the creative artist or writer in limiting the possible. There must be checks, predictions, and consequences to be tested. What fits must be consistent. Known laws of nature or scientific principles cannot be violated without providing tests for new laws. While imagination is constrained by reality, curiosity is much freer — there is no physical limit to scientific inquiry, though the methods of science serve as a corrective to that curiosity. However, there are moral limits to scientific inquiry.

## Most Scientists Live by Ethical and Moral Standards

Most scientists abide by the moral or ethical beliefs of their culture, which can be quite varied because what is legally permissible may be wider than what is morally allowed by tradition or religious upbringing. Different religions have very different moral prohibitions about what is acceptable and what is improper behavior. There is no law that forbids the artificial insemination of an ape by human sperm, and many fiction writers have speculated about the hybrid child of such an interspecific union. Most biologists would infer, based on similar crosses involving donkeys and horses, lions and tigers, and even such little known matings as D. melanogaster with D. simulans, that these crosses occasionally produce viable but often sterile offspring. Hence, it would not be a surprise to biologists that viable offspring may result. Despite the fifty or more years where such speculation occurred in print among fiction writers, no scientist has attempted such a cross. There are good reasons why they would not, and we would (properly) consider the carrying out of such an experiment either immoral or pathological in some way. Any reflective scientist would realize that if the experiment worked, someone would have to care for that offspring and it would have to be given some status as either a human being or an animal. It would create enormous legal problems in terms of whose responsibility raising that offspring would be and could lead to costly lawsuits. If the scientist worked at a university, he or she might be fired, even if tenured. This would not be seen as a case of academic freedom but as a case of moral turpitude.

Younger people who are not married and have not had much of a life outside their parent's upbringing and the schools they have attended are more likely to be scientifically curious without moral restraint. I have talked with such students who would value satisfying curiosity over the consequences of their actions, which is also why young people are more likely to experiment with drugs, sex, and other novelties in their lives than are more experienced adults. Scientists also face the same shift from relative indifference to being aware of consequences, or a more sober look at what they do and what might happen as a result of their studies. While a case like human-animal hybridization may have near universal prohibition, working on weapons that kill others clearly does not have the same moral outrage even in adult minds. Our ability to isolate morally troubling acts and legitimize them with higher values of patriotism, survival, or defense facilitates the design and manufacture of weapons. However, there is no such higher value we can use to justify a human-animal hybrid.

It is interesting that the word hybrid is derived from the Greek word *hubris*, an act that upsets the balance (suphrosne) of nature and leads to extremism that results in a damaging behavior. Thus, King Creon has the hubris of a tyrant's absolutism in his refusal to let Antigone bury her brother while Antigone has the hubris of being ruled by her passions rather than her reason. This idea of the hybrid as an imbalance was in the mind of early (pre-Mendelian) plant breeders such as Kölreuter, who interpreted the red and yellow kernels of an ear of maize from a cross of the two strains as a consequence of the "abomination against nature" that was created when two different kinds of plant were brought together. The red and yellow kernels reflected a warring of the two incompatible natures, with now one and then the other essence or nature of the warring life forms emerging. We may have abandoned such a moral interpretation for genetic crossbreeding but not for human-animal crossbreeding. Human values prevail over the generic values that applied from antiquity to relatively recent times. If we were to invoke those older universal values, almost all horticultural practices would be brought to a standstill. Scientists, like the rest of humanity, go along with a moral dualism. There are procedures that are legitimate when used for animals and plants but not for humans.

## Religious Morals Vary From one Religion to Another

Scientists are troubled by these moral issues. They have intense curiosity to know the answers to the questions they raise, but are usually held in check when the work is morally troublesome to them. I often remind them that because society tends to have a different view of what should be morally troublesome, public perceptions of science are often so damaging to science. The public sees a moral issue that the scientist ignores. It is by no means simple to decide whose moral visions should prevail. If the moral issue is primarily a religious one, such as reproductive studies by scientists who violate Church-endorsed Aristotelian natural law, non-Catholic scientists reject these imposed restrictions. Devout Catholic scientists are morally bound to accept them or face the consequences of being excommunicated if they publish their results. But if it is an issue such as germ warfare or some other unpopular activity that the public as a whole dislikes, scientists are more likely to feel some qualms of conscience if they participate in such programs, especially if there is a widespread public campaign against it. There are exceptions, of course. Those who put the government's current political values to the fore will override public protest or disapproval. I remember listening to a Canadian microbiologist give a talk about germ warfare research in Canada in 1959. He described the difficulty of producing sufficient bacteria in a given flask of nutrient culture. The bacterium he was using was *Yersinia pestis*, the agent that causes the bubonic plague. He described with pleasure his solution to the problem by sticking sterile foam plastic sponges in the flasks, and the greatly increased surface area gave immense quantities of the desired bacteria. At the time I felt an inward shudder. How could this person separate himself from the horror of using an agent that once wiped out one third of Europe's people? Did he not see the murderous consequences to non-combatant men, women, and children in the event of a war that employed germ warfare? Perhaps he rationalized that the horror was sufficient to prevent any country from using it. But it was precisely this argument that Alfred Nobel proposed when he invented dynamite. He felt that it was too horrifying in its destructiveness to be considered for war and it would be limited only to construction of civil projects by responsible engineers. Sometimes the decisions that highly educated people make are based on hope and not on wisdom, and it is luck rather than reason that protects us from our folly.

## Scientists are Governed by Restraints on Their Imaginations

Other restraints are imposed by our moral values on scientists. For instance, few scientists will experiment with their own children. Some do, but these are usually not experiments that put children at risk. Some of my colleagues in the psychology department have used their children as subjects. I recall one describing how he asked his young son to select which container had more milk, a tall thin one that was filled to the top or a larger wide container that actually had more milk but was only filled about one fifth of its height. The young boy chose the tall thin container, and the psychologist recorded this on his video recorder. Years later he had a discussion about perception with his son, who vehemently denied that he would have made such a choice and was astounded to see the replay showing that he actually chose fullness, not volume, as the measure for what is "more." J. Piaget and B. F. Skinner were among the more celebrated psychologists who utilized their children to establish the psychological theories that earned them their fame.

There is more restriction today on the use of subjects for research. Scientists have to get signed permission to do experimental work with human subjects. There is good reason why this inconvenience was imposed. In the years following the end of the Second World War, Cold War values dominated military or defense research. Experiments were conducted on human subjects who were either not informed at all that they were being tested or were misled about the real intent of the experiments for which they volunteered. Some were exposed to hazardous situations without being warned of the danger. In one experiment that a colleague of mine participated during the 1940s, LSD was put in the coffee of soldiers without their knowledge so that their aberrant behavior could be observed to see if the effects of LSD were due to suggestion or the drug itself. The suicide of one of the subjects led to a halt in the practice, although it would be difficult to prove if the suicide was related to the LSD or not.

## Abuses of Science are Still Possible Despite Moral Restraints

In another experiment carried out in California and in New York City at about the same time as the LSD study, the distributions of released bacteria were followed to see how effective germ warfare could be. The bacterium was a relatively harmless species, *Serratia marcescans*, which produces a red colony when cultured on Petri dishes. At appropriate times after the bacteria were released, its airborne presence was sampled from different places such as the subway stations of Manhattan. A few cases of *Serratia* lung infections resulted among those who had defective immune systems.

## Sometimes Secrecy is Invoked to Prevent Embarrassment to Governments

On a larger scale, thousands of soldiers were exposed to ionizing radiation during the above-ground or ground-level explosions of atomic bombs in the 1940s and 1950s in Nevada. The fallout from these tests frequently drifted over the soldiers and across Southern Utah, which did not evacuate its towns. Part of the Cold War obsession was secrecy and it was considered essential then that nothing about these tests would be made public to either our own at-risk populations, the nation, or the world. It was silly, of course, because by the 1950s, the USSR was carrying out the same kinds of secret tests on their own troops. Nothing from such secretive actions would have benefited the military of either nation, but in both cases there was a concern that public knowledge of the effects of radiation might frighten the public and make them oppose military testing.

It is not easy to impose restraints on the curiosity of scientists because restraints are profoundly influenced by culture and change from generation to generation. What is clear from the aforementioned examples is the wide range of responses to these restraints. It is doubtful, in my mind, that the taboos on human-animal crossbreeding will be broken, but I also recognize that it is not impossible that someone might do something so foolish. Even laws that forbid certain types of research do not prevent troubled individuals from breaking those laws. We also must recognize that perverse curiosities can be sanctioned by political ideology and the cover of war. Scientists are still shocked when they read of the horrible experiments carried out on Jews and other concentration camp inmates by Nazi physicians, such as immersing people in ice water, subjecting them to asphyxiating atmospheres, and burning portions of their bodies to try out skin transplants or other means to regenerate skin. Almost all scientists today agree that whatever good that might emerge from such experiments is far outweighed by the atrocity of using unwilling subjects, imperiling their health, or even causing their death.

## Scientists Seek Opportunities for Unrestrained Curiosity

Much more wholesome is the virtually unrestrained curiosity of an individual like Darwin, who spent some five years on the *Beagle* and brought back a wealth of notes, observations, specimens, and reflective findings on the geology, ecology, and biology of life in South America and island populations in the Atlantic and Pacific Oceans. When he returned, he tried out a number of experiments to test his theories. To see if seeds and branches could float from Ecuador to the Galapagos Islands, he constructed a small experimental sea water pond in which he immersed branches containing seeds for the estimated weeks or months it would take for the prevailing currents to bring them to the Galapagos. He found that many species of seeds would indeed survive their immersion in salt water and germinate. This little study illustrates beautifully how his curiosity worked. He imagined a distant event, perhaps millions of years ago, with storms tearing off branches and blowing them to sea. He also imagined these floating, larger ones serving as rafts on which small animals such as insects or birds could survive. On arrival to these volcanic islands, they would establish the first life forms. The imported vegetation, the local seaweed, and life already there from earlier invasions from other islands or the mainland would support the new animals that arrived, and they would gradually add to the organic litter to support life on the island. Darwin's tests of the branches and their seeds do not prove that this was necessarily the means by which life got its toehold on the Galapagos, but it does support the possibility that it was imported by natural means from the western shores of Ecuador. We also see Darwin's imagination at play in his book on earthworms and their role in producing soil or "vegetable mold" as he called it. He set up an area to measure the rate at which earthworms turn over the soil they are fed and convert it into organic humus with their castings (their defecation of partially digested earth). I find it not surprising at all that Darwin had such wide-ranging curiosity about little things, which is what makes scientists enjoy their work. They want to know answers and they figure out creative ways to get them from nature.

# Da Vinci's Notebooks Reveal his Remarkable Curiosity

Although I have mixed feelings about Leonardo Da Vinci because of his ease in suspending moral judgment about the consequences of his applications of science as a military engineer, I am much taken by his incredible curiosity about everything. His notebooks touch on virtually every aspect of knowledge, and he is astronomer, geologist, biologist, chemist, and physicist all in one. He used science to invent new techniques for his frescoes and paintings, and he devised ways to cast sculpting that looked inherently unstable. He was as attracted to human behavioral differences as to the varieties of animal behavior, and he also looked with longing on birds, studying their flight and learning how they glide and stay in the air. Years later, like Darwin, Da Vinci recognized that the presence of shellfish fossils in the hills and mountains of Italy reflect a past submersion of that land under water.

The passion to learn drives science. The most creative scientists exercise their imaginations and learn not to blunt their curiosity within the realms of inquiry that are not morally offensive. A Freudian psychiatrist may attribute such a focus, drive, or passion to know about things as some sublimated sexual curiosity, but we are unlikely to know the exact sources of the scientist's zeal for knowledge. It is also not unique to science because almost all bright children and youths have this trait in abundance. I have read thousands of applications to medical school and to the Honors College at Stony Brook, and I have served on many committees that provide awards and honors to students. I am both pleased and constantly surprised at the activities that appeal to our most talented high school students, who have a curiosity about so many fields and enjoy working in the arts, music, literature, science, mathematics, history, and sports. It is as if the young body and mind want to plunge into everything there is to know, sample everything, and celebrate each new piece of the universe that is discovered. For many a scientist already in old age, the constant immersion in curiosity and imagination required by science as a discipline keeps the scientist mentally alive and provides an illusion of perpetual youth.

# 19 The Vulnerability of Science

I do not know how much of the general public fears, hates, or loves science. I am also certainly not out to make all of humanity love science — this will never happen just like how it would be impossible for everyone to love opera, poetry, or ballet. People's tastes will always vary and they will have likes and dislikes. I can understand that personal taste; for example, I have dislikes for some fields of knowledge, including economics, business, and theology. If I tried to like these areas, I suppose I could learn to appreciate them, and I have had brief but not happy excursions into these fields. Such preferences are not too different from food prejudices — I have yet to host a dinner party for students without encountering at least one student who hates tomatoes, another mushrooms or eggplant, and yet another onions, and on it goes. I totally expect it; I have had far more food prejudices as a teenager and young man than I do now.

## Some Fear Science as a Personal Threat to Their Religious Beliefs

I am more concerned about a deep fear of science as something that corrupts minds or reveals them as incompetent. I also feel sorry that there are people who hate all science and not just thoughtless or bad science. These are some of the people whom I hope will learn more about what science is and what some of its positive virtues are. For some of those who love science, I wish that they stood back sometimes and reflected on their values. All three reactions — love, hate, and fear of science — make science vulnerable.

I will consider the vulnerabilities and responses I would make to each of these three camps. Let us begin with religion. For a long time, science and religion have been at war with intervals of armistice but never peace. It is difficult for the two to feel fully comfortable with each other. For those whose religion includes a good deal of pseudoscience, inaccurate representations of the material world, and a demand that its members put faith over reason on matters that can be readily tested, there will be no end to that war. There are significant portions of the population in the United States (but probably not a majority) who feel strongly that the Bible is inerrant and that their particular religious denomination has an accurate reading of what it says. This makes evolution play a more intense role than other aspects of science in their disappointment. That dislike of evolution is not limited to biology because at stake is the age of the universe and earth, which astronomers, physicists, and geologists have also assigned as millions or billions of years. A far larger proportion of Americans, mostly Protestants, feel uneasy about evolution but do not automatically reject it. Many of them compromise and see evolution as God's way of making life. Frequently this evolutionary perspective is seen as purposive and driven or guided along the way by God. This is not God the winder of the clockwork universe who sits back and observes nature's laws in action, but instead a God who is believed in and prayed to, and who makes his presence known through the personal religious experiences that so many churchgoers enjoy. Most Catholics and Jews are not troubled by evolution.

## Some Fear Science as Eroding Their Children's Faiths

For those who feel uneasy, there is a legitimate concern that science often overlooks. When children are confronted by the science they learn in school and the religious views they learn in church, they quickly realize that they are incompatible. The earth cannot simultaneously be 10,000 years old *and* 4.5 billion years old. Life cannot have been put on earth as it looks now *and* at the same time have evolved gradually with virtually none of today's species being on earth hundreds of millions of years ago. Adam could not have been the first human while his species, *Homo sapiens*, is only the latest of several, now extinct, species of *Homo*. Parents and their religious community rightfully feel that their child's faith will be weakened if they are exposed to such thoughts. Many (not just scientists) argue that a faith is as good as its persuasive power. If the faith is so full of holes that it can only be shored up in isolation from the world of knowledge, then it is a weak faith to begin with. The accusation against science also used to be made against any new religion entering the community. People in one religion do not like to see their members converted to another religion. This resistance has, however, never stopped missionaries from their evangelistic efforts.

There is another feature of this strain between religion and science that is worth noting and has nothing to do with the theology of creation. Science, in general, is a secular endeavor or occupation, and it is also very successful and influential. Young people tend to be more skeptical of their heritage and at college age they are more likely to stop going to churches or synagogues and consider themselves non-observant. Few in the United States are willing to call themselves atheists because this is still disapproved in our culture. Atheism conveys a militant stance against a belief in God, which is no doubt how some atheists see themselves. But atheism is also about living a moral life without the need for a god concept. The public, however, does not always make such fine distinctions and the pugnacious atheist is more typically perceived when a young adult claims to be an atheist. Thus for a period of time, college-aged students to a large degree identify as neither atheist nor observant. They enjoy the pleasures of the secular world, feel uncomfortable about their legacy of beliefs which they have largely shed, and may even consider their relatives hypocrites for professing a faith while acting differently from its ideals. Generational intolerance and youth frequently go together. Alternatively, we can say that youths value consistency in moral and ethical beliefs among those who raised them. They become disillusioned when they see the complexities of adult lives, which are all too real, and the compromises that adults must constantly make. Science is often blamed for seducing the young into secular life. I suspect that long before there was a well-defined science, there were other secular temptations that similarly appealed to the young — the worlds of money, sex, art, athletics, and power are much older than the world of science. Older adults have learned to deal with these secular features of society but not science because it is too new.

## Some Hate Science Because it Challenges Their Biases

Science is also vulnerable to attack from those who live in ignorance. Here, the image of the Luddite sends a chill through science. After Joseph Priestley's laboratory was burned, he finally emigrated to North America to escape persecution for his religious heresy (his Unitarianism) and his strange chemical experiments (he discovered oxygen), which looked to the ignorant like he was practicing some diabolical arts. Ignorance is the companion of bigotry, racism, superstition, and pseudoscience. They feed on ignorance because there is no strong challenge to spurious views. Science is good at exploding lies and self-deception, but because people like to believe that they are not being deceived by others, they are sometimes more angry at the messenger who reveals that they have been duped than they are by the charlatan who dupes them. I can understand that. It hurts to put a lot of faith in a person or community that is supportive and makes you feel good about yourself only to learn that they are deceptive or untrue. Too often the scientist plays the role of illusion-buster. The scientist refutes superstition with facts and a materialistic reality that demolishes cherished beliefs. This may make a community unhappy and science would be seen as a foe of society. The scientist, however, is not a healer of souls and cannot provide a wholesome replacement for a shattered belief system, whether it is a fake faith healer, a psychic surgeon, a palm reader, or a spiritualist claiming to bring back dead relatives. Cults are vulnerable to such letdowns when their leaders are arrested for tax evasion or other shady practices that violate civil law. Many parents, themselves observant in their beliefs, fear cults and are worried to learn that a son or daughter has gotten involved with one.

## Science Challenges Pseudoscience and Speaks out Against it

While scientists tend to avoid confronting religions, they will take on pseudoscience and expose the fallacies used by astrologers, medical quacks, circle-squarers, inventors of perpetual motion machines, and others who dupe the gullible (including poorly informed legislators and financial investors). Science risks being looked on as a bullying, dogmatic Goliath to those who feel like modest Davids, who are only hoping to demonstrate a new law of science or some process that defies traditional rules of entropy. Those who practice pseudoscience and their supporters see science as an elitist club that excludes them and demands orthodox belief as the price of membership. The public often believes the person who lays claim to being hounded. Some of their supporters may feel that science is right but nonetheless still wrong for taking the fun out of life. People should indulge in some fantasy, they say. Isn't that why people put a few dollars on lottery tickets hoping that birth years and other significant numbers will be the winning combination to millions of dollars? True, some people like to have their fortunes told by a medium despite knowing that it is fake. I have a friend who delights in buying counterfeit wristwatches. He knows they are fake, but he considers it a way to dilute the snobbishness of those who can afford the real thing. Some look on the pseudosciences as enjoyable, like reading science fiction or fantasy, and it gives them a feeling of significance that science denies. Why be a nobody when you can be made to feel like someone very special?

## Some People Prefer Illusions to Reality

I am not sure how a gravely ill person benefits when they consult quacks who promise treatments that are actually worthless. They may save money, but it is not money that they can use when they are dead. Does a person who knows that nothing will work benefit by just resigning oneself to death? Sometimes an illusion keeps people going because it adds hope to their lives. I am sympathetic to such persons, but I also resent the greed of the quack who knows that the treatments are phony and money is gained out of another person's desperation to avoid death. It is difficult to think of an effective alternative for the quack's victims. But when a quack promises a cure for a child's retinoblastoma and offers prayer and faith as the only successful means to treat the cancer, that is an immoral act if not a criminal one. Some 95 percent of children with retinoblastoma can be treated successfully if the condition is caught early. If religious faith makes the parents choose the healer over the physician, this is an occasion to call the healer a fake.

## It is not Easy for Scientists to Function when the State Imposes Ideological Restraints

The relationship between science and the state has always been troubled. The extremes are the government-endorsed ideologies that masquerade as legitimate science. We saw what Lysenkoism in the USSR and social and medical science in Nazi Germany were like when the governments supporting them had total control. People living in a democracy feel protected by Constitutional guarantees, and science is less vulnerable to censorship or ideological manipulation than elsewhere. However, science is not free of oversight for funding. Work that is too controversial or sacrilegious would have difficulty being awarded federal grants. There were many tests of loyalty during the Cold War, and students who received federal fellowships to study science had to sign loyalty oaths. Communists and their sympathizers were forced out of many universities. Even persons of conscience who felt that loyalty oaths were demeaning were forced to resign. The debates cut both ways. Muller had little sympathy for teachers who alleged that they were Lysenkoists not because they were Communist but because they claimed that Lysenkoism was an infant science that had a right to be taught so that students could weigh its merits. He was convinced that they should be treated as ideologues, the way that Nazi professors were regarded in the late 1930s when they advocated, in their classes, Aryan supremacy and the inferiority of Jews and other races. Academic freedom frequently lost its meaning when it was asked to support doctrines that had contempt for scholarship, objectivity, and the integrity of science. It was as difficult a time for those who believed in their ideologies as for those who believed that they had to protect intellectual honesty or those who believed that no ideology, however wrong, should be barred from the classroom. Political reformers, subversives, radicals, and reactionaries are as likely to participate in scientific fields as they are elsewhere in the academic world.

# Numerous Groups or Causes Seek Scientific Support and Sensitivity to Their Needs

Special interest groups appear from time to time and bring their causes to bear on science, each with its own way of wanting scientific support while wanting science purged of some of its past or alleged defects. The physically or mentally handicapped, disabled, or challenged want non-wounding terms to replace older ones that were or have become demeaning. Medical science readily complied. A similar purging of habits and names took place in response to the feminist movement. The outcomes are not fully consistent. For instance, the term "venereal disease" was abandoned because of the implication that women are the cause (via Venus, the goddess of love) and was replaced by "sexually transmitted diseases." Hence, VD became STD. But masturbation still stands although its etymology includes *manus* (hand) and *strupare* (to defile). The sin of Onan still lives in masturbation. On the whole, most of these changes are for the better. Down syndrome is an improvement from mongoloid idiocy, and Hurler syndrome is better than gargoylism. Some terms have become awkward or silly. When "exceptional child" replaced retarded, idiot, imbecile, or feebleminded, the term is so neutral that it does not really convey the problem. Mentally slow or retarded is not necessarily pejorative. For those who wish to deny that the child has difficulty learning, exceptional child is perhaps fine, but it only satisfies the parents or those who find departures from normal learning uncomfortable to acknowledge. An eponym would be much better. For example, I would offer "Penrose condition" for children whose academic performance is impaired due to genetic, environmental, or social reasons. Lionel Penrose was the first physician to recognize that mental retardation is highly heterogeneous and that it is not easy to identify specific causes for learning disabilities. Exceptional child suggests ideological motive or simple denial as the reason for its construction.

Over the years, I have learned how to construct sentences to avoid gender bias. I think that this is a good change. I use the plural where possible to avoid the "his or her" trap, and I avoid, as much as possible, the use of "one" to refer to a person. Language constantly changes and all we need to do if we want to see how phrasings have changed or how words have taken on new meanings is to go back to Victorian or Elizabethan times. Scientific language has also gone through changes. In the genetics literature from the 1890s to the end of World War I, the language was frequently so personal and polemic that you would know what bothered an author. When William Castle got irritated by one of Morgan's students calling him to task for conceptual errors that allegedly smacked of "mysticism," he responded with a paper titled "Mr. Muller on the constancy of Mendelian factors." After that war, someone changed the rules of science publication. Ad hominem comments, venting spleen in print, and first person writing went out. Science became impersonal and in the third person, and a lazy habit of inactive construction came in. Science writing became boring and acquired a flat technical style of expression that closed it off to readers other than fellow scientists. When I was an undergraduate at NYU, I had written a laboratory report for my freshman chemistry class in the active voice and used a few adjectives to make the sentences come alive. The teaching assistant put me down with a quote, "And as the sun sets among the palm trees in lovely Bermuda..." telling me that cutesy comprehensible English was inappropriate and that good science had to be written in the third person passive. In the 1970s, the writing was so bad that editors began to reverse the trend. Titles started becoming complete sentences, the first person reappeared in some articles, and the active voice was brought back into sentence structure.

## Science is Sometimes Used to Bolster Questionable Claims

There is concern over attempts to impose poor scholarship and science as legitimate science when it justifies an ideology. This is occasionally seen in the disputes of those who seek a cultural or racial stamp on science. I cannot estimate how serious these efforts are in raising ethnic and racial consciousness mainly because most scientists ignore these efforts while those who advocate them do not or cannot publish in mainstream journals read by most scientists. I have read some articles claiming that melanin is intrinsically good and is in some causal way a basis for behavioral traits that are beneficial to Africans and African Americans. By contrast, the paucity of melanin in whites (Caucasians or "ice people" as they were described) deprives them of these beneficial behaviors. This view is entirely different from those espoused by more traditional scientific studies of melanin, which show that melanin is adaptive in screening ultraviolet light and lessens the risk of skin cancer and life-threatening melanomas. For this reason, African Americans are not as much at risk to such cancers. But light skin color in Northern latitudes permits the skin to get enough ultraviolet to convert cholesterol to vitamin D and thus protects children from rickets. Dark-skinned individuals in such latitudes would be more vulnerable to having children with rickets. I certainly teach these ideas on the relation of melanin and skin color to these very different consequences of ultraviolet on the skin. The mechanisms are quite different. The absorption of ultraviolet by DNA leads to thymine dimer formation and the opportunity for mutations and chromosome rearrangements during repair, while ultraviolet provides energy for the synthesis of vitamin D from cholesterol. It makes no sense to speak of these in terms of racial superiority. Adaptation is a legitimate term when discussing human evolution; superiority or inferiority of a class of people always smacks of racism or unexamined bias. Those who embrace these doctrines need to be reminded of similar racial theories advocated by white supremacists and of nationalistic excess in the Soviet attempt to rewrite the history of science with a Marxist or Slavic style in reaction to their view that Western science was tainted by capitalist, idealist, or crudely mechanistic biases.

I have discussed the misgivings that natural scientists feel when reading the critiques of science by historians, philosophers, and sociologists who see science as a construction instead of a description of reality. Most scientists are weak in these three areas because they are immersed in doing what they love, which is describing reality. Telling scientists that they are just constructing some model, paradigm, or consensual worldview has little effect on the way scientists do science. They will ignore such criticisms as poorly informed. I have tried to provide those who belong to the constructivist camp a richer and more detailed picture of what science is. It may well be that those in the history, philosophy, and sociology of science are likewise impervious to criticism of a root belief, and it may also be true that the "two cultures" described by C. P. Snow no longer reflect science vs the arts and humanities but rather the natural sciences vs the social sciences.

## Scientists Reject Constructivist Views of Reality as Non-scientific

Most of my colleagues who are interested in discussing this issue do not fear that scientists will defect to the constructivist model. Instead, they fear that new generations of college students will be convinced by teachers in these fields that science is just another arbitrary enterprise that depends on mutual admiration and power and thus deserves no more support from taxpayers than the local Chamber of Commerce or the United Federation of Teachers. I try to remind myself that the public goes with success. Constructed and arbitrary realities do not yield reliable results; science does. When there are public needs for the applications of science, these constructivist models will be ignored.

The public fear of science is not unreasonable. There are indifferent corporations that do not pay much heed to worker safety or public safety unless they are forced into such behavior by federal or state regulation. Despite the cries of those who believe that business would regulate itself and that the government should not interfere with free trade or the decisions that corporations must make to survive and grow, the public is generally supportive of these regulations. It is they who have experienced slums, urban blight, sprawling malls and shopping corridors, the uglification of streets along railroads, the ghastly odors of garbage dumps, the smell of bromine and the discoloration of air around chemical industries, and the disappearance of parks, woodlands, and clean rivers.

## The Role of Science in Military Development Remains Controversial

The public is less active in questioning the sums of money that go to military preparedness. Patriotism runs deep in any country and few wish to quarrel with their military advisors or governments on the real or imagined threats faced or what is needed for defense against aggression. It is the one area in a democracy where the experts have their run of influence. By contrast, militarism was considered lower in priority during the Great Depression and elected officials were held accountable and had to justify the tax moneys that went to the military. Some of my teachers remarked during my elementary school classes that what made America so wonderful for immigrants was that they did not have to participate in obligate military duty. These were not radicals or America-first crypto-fascists; they were school teachers who made us pledge ourselves to the flag and took a great deal of pride in being American.

Public fears of science running amok with new technologies is a recurrent theme. Scientists have learned to organize and discuss the potential troubles of their findings. We have the leadership role of *The Bulletin of Atomic Scientists*, which informed not only physicists but all scientists of the ethical and political issues scientists faced doing science during the Cold War. Their warnings rubbed off on molecular biologists and it was admirable that scientists themselves were the first to make a public statement about their concerns over the future of recombinant DNA. A similar effort is underway for the future uses of the human genome project. Although that fear is often stated as science going "too far, too soon," most new fields of science and new technologies have encountered similar views in the past. I consider this public fear of science healthy as it pushes scientists to think carefully about the consequences of what they do. The public has a right to ask this of the scientists whose work they support through their taxes.

# 20 The Complexity of Science

Normally, science simplifies our understanding of the universe. Heredity and variation ceased being mysteries and became comprehensible through Mendelian laws, the chromosome theory of heredity, the theory of the gene, and the molecular basis of gene structure and function. What was murky at the end of the nineteenth century became clarified in the twentieth century. Less well understood is a complex science like weather forecasting. In the past, seamen often used experience to assess the weather and when my father was working as a seaman on a cargo ship, he was taught "red sky at night, sailor's delight; red sky in the morning, sailor's warning." Roughly, that saying held. But sometimes science remains complex and difficult to sort out into laws, predictability, or clear principles.

Weather forecasting at the start of the nineteenth century was largely dependent on weather stations, a concept introduced by Humboldt in the 1840s, augmented by Mendel in the 1860s, and substantially increased by Galton in the late nineteenth century. By that time, it was possible to send information by cable around the world and this made prediction better. Yet even in the 1930s, major hurricanes were missed and devastation could not be avoided along the east coast of the United States or the Gulf coast. By the first decade of the twenty-first century, such disasters became less likely because weather patterns can be discerned by space satellites and hurricanes are tracked from their start as depressions and tropical squalls to gathering hurricanes with reasonably predicted paths the closer they come to the United States from their origins off the African or Mediterranean coasts. Many features are still uncertain, such as the category assigned to the destructiveness of hurricanes. Predicted Category 5 storms often become Category 2 storms when they hit land. Some weather patterns, like the durations of droughts, cannot be easily predicted.

## Complex Science is Associated with Increased Numbers of Variables

Complexity arises in weather systems because there are many variables such as ocean currents, wind direction, and temperature. Computers are required to digest immense volumes of past histories of storms, tornadoes, hurricanes, droughts, prolonged rainfalls, and other data of concern to farmers, travelers, and those in the path of destructive weather. The capacity to predict improves with each generation, but it is still an imperfect science because so many variables are involved. Less clear is the resolution of controversies over the ecological and health effects of pollutants as well as the effects they may have on climate, such as acid rain or bad smog days. The consequences of long term use of hydrocarbons for fuel and their relation to climate change are similarly complex. Unlike weather forecasting, which is rarely controversial unless a major storm is somehow missed or misclassified in its predicted damage, the environmentally associated activities of carbon dioxide formation, soot, ozone, sulfuric acid-forming compounds, nitric acid-forming compounds, synthetic agents used as refrigerants, pesticide and herbicide usage on a large commercial scale, arsenates and other wastes from manufacturing, and similar concerns are always controversial. A good deal of the attention given by environmentalists in the twentieth century to concerns over pollution have shifted in the twenty-first century to a concern over climate change, often seen by the public as a concern over global warming. Those who have raised the concern have noted annual increases in the mean temperature of the world using United Nations reports. These reports have also noted the melting of glaciers and a rise in sea levels throughout the world. The arctic has become substantially ice free in the summer, opening up a "Northwest Passage" for commercial passage. Some species have moved north, changing the ecological status of the northern niches. People who see climate change as largely a product of human dependency on fossil fuels also point out the huge increase in carbon dioxide in the atmosphere. Critics of climate change argue that the climate has natural cycles and these may be a result of changes in the sun's output, natural shifts associated with the earth's wobble around its axis, ocean current shifts, or major changes in microbial metabolism.

The most obvious reason for these controversies is that the products or industrial activities accused of polluting or contributing to climate change are of considerable commercial value and involve jobs, corporate profits, and sometimes military needs. That shifts them from the curiosity of scientists trying to understand why things are changing and puts the findings in conflict with those who can influence legislation. That legislation is often regulatory. Manufacturers would prefer no regulation at all in how products are made, used, or disposed of. Environmental scientists frequently lend their support to both sides of these conflicts, and resolving them is very difficult. Good science is rare when issues are complex because the clear-cut answers desired by legislators, journalists, or voters do not exist. It is obvious why this is so if I were to ask a meteorologist what the weather will be like six months from now in the community where you live — it would be largely guesswork hedged with probabilities. A geneticist could give a good estimate of mutation rates if the sperm of fruit flies were exposed to, say, 1,000 roentgens, but that exactness becomes more difficult to attain as the dose drops to 200 roentgens, ten roentgens, or that of a single chest x-ray. It is also unsurprising that those who consider low dose exposures harmless (or even beneficial) are often those who come from industries that use radiation or are supported by agencies or private philanthropies that are heavily involved in using radiation for industrial, military, or health uses.

## Designing Experiments for Complex Science is Difficult or Impossible

I was very frustrated when I worked in environmental mutagenesis because of the difficulties I faced trying to design experiments that would reveal the safety or harm of very low doses of radiation or chemicals. In fact, I discovered many of the reasons why good experiments are difficult to do. If I injected fruit fly males with a potent chemical mutagen like quinacrine mustard, I could get lots of mutations and work out how they differed from those induced by x-rays. But if I used agents like LSD (lysergic acid diethylamide) even at several hundred times the dose that is used in a typical human trip, they had no effect on mutations, chromosome breaks, or gains or losses of chromosomes during cell division. This finding arose despite the similarity in structure of LSD to quinacrine mustard. Adding reactive chlorines to the side groups of a complex molecule can shift its activity to that of a potent mutagen. But if I injected female fruit flies with the same high dose of quinacrine mustard, the mutation rate was dramatically lower. I believe that this outcome occurs because the mutagen bathes the sperm where the DNA of the chromosomes is close to the cell membrane and virtually no cytoplasm is present, whereas that DNA in an egg is shielded by a huge wad of cytoplasm around the nucleus and thus little of the chemical gets to the DNA.

If we shift the problem to humans, I would be concerned about lung cancers if the chemical were volatile and inhaled daily in the workplace (like formaldehyde fumes in cloth manufacture). If it were dissolved in drinking water as a contaminant from manufacturing discharge, I would look at the potential for cancers of the liver, kidneys, or bladder. If it were likely to remain in the intestine from digested foods or swallowed products like tobacco juice from chewing tobacco, smoking pipes, cigars, or cigarettes, I would also look at throat cancer, esophageal cancer, and intestinal cancers. Simple answers are difficult to obtain when the sites of damage are diverse and related to different agents and how they act in different tissues. Such confounding factors also make it difficult to give answers in court when yes or no replies are expected and opportunities for mini-lectures are squelched.

## Agent Orange Effects Remain Difficult to Assess

I ran into the same problem while studying the effects of Agent Orange. Agent Orange is a mixture of two herbicides, 2,4-D and 2,4,5-T. Only 2,4,5-T when manufactured produces a compound called dioxin, which is a very toxic chemical. Thus, there were three components in Agent Orange during the early stages of the Vietnam War. When dioxin was found to be the most dangerous of these three, the chemical manufacture of 2,4,5-T shifted to using lower temperatures and hence lower amounts of dioxin. But both 2,4-D and 2,4,5-T have developmental effects in fruit flies. When they were shown to also produce developmental effects in small mammals, they were phased out from use in Vietnam. At issue in the Agent

Orange controversy are the following problems. How much damage was done to the forests sprayed and the variety of species in those areas, and what ecological changes occurred after the war? How effective was Agent Orange in revealing enemy supplies, camps, and trails? How effective was defoliation as a weapon to diminish enemy combat? How effective was Agent Orange in destroying enemy crops? Were there health hazards for those sprayed or living in sprayed areas? Were the effects different for the Vietnamese living there than for the allied forces passing through? I spent a lot of time reading the literature on this topic and my assessment left me with contradictory views. Chemical companies, of course, denied health effects and when they occurred claimed that their product was misused by the military. The military was divided; some officers thought that it was effective in clearing vegetation and thus reducing ambush. Others claimed that it had virtually no military value. Some surveys by public health services and physicians reported higher incidences of illnesses in returning veterans while others did not. Confounding the science were the politics that went with these controversies. American veterans of that war, when they had children with birth defects, claimed that their exposure to Agent Orange was the source of their children's medical conditions. But how do you prove this? Birth defects normally involve about 3 to 5 percent of all births depending on what is considered as a birth defect. Low dose exposures (for the overwhelming number of veterans) are at least as complex for Agent Orange as for radiation exposure, but virtually no veterans exposed to radiation in the 1940s and 1950s during Cold War tests of atomic bombs in Utah and Nevada argued that their children's birth defects were radiation-induced. This was partly because the induction of birth defects was unknown to the public until the thalidomide disaster of 1958–1962 when an over-the-counter drug in Germany was used, mostly in Europe and the United Kingdom, resulting in 8,000 babies born with deformed limbs and other serious birth defects.

## Mid-course Corrections Solved Some Complexity Problems for Space Science

Sometimes complex science can be resolved by a different approach. When the space program began, there was concern that there were too many variables involved in programming a rocket to reach the moon, photograph it while in orbit, and return to earth. Instead engineers used "mid-course corrections" where departures from trajectory could be slowed down, speeded up, or directed along new paths by commands from earth-based computers that fed data into the rockets' computers and allowed these adjustments to take place. Note that this was highly successful even though the initial trajectory could not be programmed with a high level of accuracy as the gravities of the earth, sun, and moon interacted while the rocket moved away from the earth.

Difficulties in applied science may be overcome by getting many scientists involved and ensuring that there is abundant funding. This approach worked well for the development of the radar during World War II and the production of atomic weapons in the Manhattan Project. But similarly large sums of money and large numbers of funded scientists have not led to a breakthrough in "the war on cancer" which was declared by President Nixon a half century ago. Unlike mid-course corrections for space flights, no such correction exists for combating most cancers. There are a few cancers whose incidence has been reduced and there is some prolongation of life with chemotherapy and other approaches to destroy cancer cells, but virtually no insights exist to prevent tumor cell formation for most of the cancers that humans experience (e.g., breast, prostate, colorectal, leukemia), except for lung cancers where the overwhelming number are associated with tobacco smoke inhalation. Contrast cancer with the elimination of smallpox and polio. Public health officials were able to use immunizations to prevent the spread of these diseases and target pockets of people who were still not immunized against these diseases. The absence of animal vectors to serve as alternative hosts to the virus also made these diseases easier to eliminate. Acquired immunodeficiency syndrome (AIDS) arose in epidemic form in the 1980s and rapidly spread to most of the world, especially to vulnerable populations such as homosexual men, intravenous drug users, and those engaging in unprotected sex (mostly not using condoms). Complicating the AIDS story were religious prohibitions against sexual activity in unmarried couples, religious and social prejudices against homosexuality, the criminal association of drug users using hypodermic needles, and the political climate where many people felt that those who came down with AIDS were victims of their own sins or crimes. Unlike polio and smallpox,

HIV infection was seen by some as a punishment. From the scientist's perspective, AIDS is the story of the human immunodeficiency virus (HIV), which can be prevented by sexual prevention (safe sex, abstinence, or both), using sterile rather than shared hypodermic syringes in the case of intravenous drug use, and treatment with a variety of agents that damage viral DNA or viral proteins. In the United States, religious demand for celibate behavior and noncriminal behavior (e.g., "Just say no") has been favored at the government level because of the lobbying effort by religious leaders who see AIDS in moral terms rather than as a pathology of infection. In this case, the complexity is primarily due to political and social factors rather an intrinsic difficulty in how the virus spreads. One unfortunate aspect of the AIDS epidemic is the lack of an effective vaccine. Many attempts have been made to find one, but HIV has turned out to be a complicated virus because it infects the cells of the immune system and has developed numerous ways to dodge natural immunity responses in humans. In that one aspect, the HIV infection allows us to think of the science involved in its conquest as complex. Fortunately, scientists have worked out effective medications that can keep the virus level so low that HIV positive persons cannot transmit the virus to their partners, and those who are pregnant cannot transmit the virus to their embryos or fetuses.

What should our response be to complex science that does not offer easy answers to those who demand rapid and straightforward responses from scientists? One approach would be to boost the public education and awareness of science. As long as we simplify our presentation of science and students see it as simply some controlled experiment with a single variable under watch, students will recognize that science is a viable way to handle any problem. Much harder is to teach students at the high school level that sometimes the issue is not the science itself but government, industry, religion, or patriotism (in times of war or preparation for future wars) that play roles in making the science more complex than it should be. Still other times the scientific issue itself involves so many variables that no amount of government money can produce solutions in a short time. Both are important to learn, but they are also controversial and some solutions to these complex problems may require numerous generations before they are resolved. Fortunately, society can shift its values, allowing opportunities for science not encumbered by such restraints to work out solutions.

# 21 What is Science?

There are many sciences and they share in common an understanding of some aspect of the universe. How they get that understanding and the tools they use to achieve assurance in objectivity varies with the science. The classification of sciences has a certain arbitrary aspect to it because the sciences overlap and feed into each other. Some of the classifications reflect bias on what is inside the scientific realm and what is outside. I have never liked the older classification of exact (physical) and inexact (biological and behavioral) sciences. It may have made sense a century ago when the physical world seemed more contained, but the very small (e.g., subatomic particles) and the very large (e.g., black holes, singularities) have made it difficult to call physics an exact science. All sciences, in fact, are inexact although for different reasons. They study components that are not readily accessible and depend on theories that hold together isolated bits of information that can yield more than one interpretation.

## The Major Branches of Science are Like Continents

I like to think of the sciences like continents scattered around the earth. There is one large area we call physics and another called astronomy, and these are related through an intellectual traffic going back and forth called cosmology. Astronomy and the area we call geology are connected through their mutual interest in the solar system and the planetary sciences. We have an emerging field of close-up studies of the planets and their satellites, and we have already landed instruments giving us local information on

Venus, Mars, and the moon. In a century or so, all the major planets and their satellites will likely have had similar landings to relay information from sophisticated robots. Physics and the area called chemistry are connected through studies of crystals among other physical properties of matter, but this is only one of several mutual interests that chemists and physicists enjoy. Both geology and chemistry have a lot of shared interests with biology. There is biochemistry, molecular biology, and the study of macromolecules that cover the chemistry-biology flank as well as ecology and evolution covering the geology-biology flank. Biology in turn is intimately connected to the social sciences. Anthropology, psychology, sociology, political science, economics, and history all have common interests with biology through human biology and medicine.

Outside these continents of science and the free-flowing traffic that connects them to varying degrees are the areas that science excludes. Using our earthly analogy, it is much like the coalitions of nations that formed different alliances in the nineteenth and twentieth centuries. The analogy may fail when we compare the way alliances realign and old values fall out of favor as new states and new interests predominate. The sciences have tended to keep together for a much longer time than individual nations in their blocs. But branches that once were accepted as science were purged, most of them long ago. Alchemy got kicked out of chemistry, astrology got out of astronomy, and minor fields like phrenology, physiognomy, and parapsychology have largely been demoted from the life sciences. Religion (in its supernatural claims) and kindred spirit worlds have been expelled from the continents of science. They make forays back and forth but there is no mutual acceptance as there is with geology and biology or with physics and chemistry.

# The Methodologies of Science Differ

Each of these large fields of science has its own methodologies. I was struck by that a number of years ago when I was looking at papers in genetics. They had no pictures, but lots of tables of data and descriptions of how experiments were done. However, all the reprints I had seen in the field of embryology had photographs and drawings. It was not enough to describe a procedure for switching bits of tissue between embryos; photos were needed to firm up the evidence and give it authenticity. I cannot think of a geneticist photographing, let us say, a porcelain plate of separated flies to show the evidence of a particular result from breeding analysis. Some fields emphasize pure mathematics while others emphasize applied mathematics, especially statistics. Some are heavily experimental and doing laboratory work is considered the only legitimate way to do the science. Some rely mostly on observation (astronomers have a difficult time altering the conditions of things outside the solar system). Others require heavy investment of time in field work, like geology or its cognate science, oceanography. The instruments they use vary. Astronomy needs telescopes and rarely microscopes, but biology relies heavily on microscopes and rarely telescopes. Chromatography would be essential in chemistry but may play little or no role in astronomy. In general, the social sciences use few instruments. Physics depends on very costly instruments.

I gave a talk once at the Fermi lab in Illinois and I was given a tour of the tunnels housing the cyclotron. It is immense. I learned from my host that one experiment may employ several hundred PhDs to work on setting up magnets and detectors in the great rings of wires that are wrapped around the cyclotron. Testing each of these units occupies a lot of time. It may take one year to prepare the cyclotron for a given experiment, only one or a few days to run the experiment, and then another year for the team to divvy up the data and look at tens of thousands of photographs of colliding atoms to determine if a rare particle has been caught in the process of forming or decaying. A single experiment of this magnitude may cost several million dollars. The scientist in charge of the experiment is often like an army general, keeping track of many people who each have a part to play in a huge project that must be meticulously coordinated. Error is too costly for details to be left unplanned.

## Science Depends on its Integrity

If science consists of many fields, each with its own way of getting information and adding to our understanding, a second feature of science is its reliance on validation or truth. Because science cannot work with false or unreliable data, data must be accurately described and how that data was obtained must be clear so that anyone can repeat the procedures and see if the authors of the article have indeed made a contribution. I read several years ago about a geologist in India who faked his findings by buying fossils and planting them in appropriate hills where he and his team would "discover" them. He was unfrocked because he erred in assuming that the unusual locations for his specimens would not be noticed. His specimens were examined and they could not disguise the tell-tale chemical and physical evidence that they had an origin other than the hills where they were allegedly found. I share with Kant great reverence for the truth, so I virtually never lie because of my respect for Kant's claim that a lie distorts our sense of reality and makes it difficult for us to reason or know what is true. Fortunately, when scientists check the data of other experiments, they have independent means of checking its provenance and legitimacy. One of the reasons that studies in parapsychology and intelligence testing have fallen in such disrepute is the history of fraud in these emotionally charged fields. Some of the chief advocates of these fields were themselves caught using fraudulent practices or did not protect their data from the contamination of selective bias. What we expect and what we get may be quite different. Most scientists take great pains to avoid being caught deceiving themselves by their expectations. At times some scholars ridicule the notion of truth and equate it with some cosmic revelation that may be a big T truth. Quite often the simple little t truth of science makes these grandiose claims of truth no longer achievable in the real world. If a scientist counts so many flies and separates them into categories and enters the data, there is a presumption that the data are as seen and no conscious effort was made to construct the data (i.e., fake it). I remember the anger I felt at one of my classmates when I was an undergraduate taking a genetics laboratory course at NYU. I came in on weekends and odd hours to count my flies and enter the data. One classmate used a slide rule to estimate what he should have seen, used that as evidence for his paper, and used his time to pursue whatever other priorities that he had put ahead of the truth. He told me that I was a fool because the professor did not care, as was evidenced by his almost never coming to the laboratory to help us or to give us advice. Scientists who care about reality and their attempts to understand it will not let such social factors undermine their primary values. This is why they are scientists. I remember Irwin Herskowitz in Muller's laboratory once telling the graduate students after a case of fraud was circulating the campuses, "Never feel sorry for me. Never give me the data that I want. Only give me the data you get."

## Science Believes it is Describing Nature

Scientists have a respect for nature. They are confident that there is a reality that can be described. It may not be fully described and one could also argue successfully that nothing is fully capable of being described because our asking of impossible questions will get us no answers. You could give me a crystal of table salt and ask me how many atoms are in it, and I could give an approximation but not an exact count. I am not sure that anyone could; there may be no instrumentation available that is so precise as to count every single sodium and chlorine atom in the crystal. In fact, such a demand would be unfair. If a scientist can give an estimate of how many sodium atoms are present in a tablespoon of salt, let us say about ten to the $22^{nd}$ power, this would be sufficient. If the basic information on its composition, sodium and chlorine, is given and the organization of these elements in the crystal (a simple right angled lattice) is provided, this is a fairly decent description of the crystal. Lots more answers can be produced as one discusses the other properties of the crystal (e.g., its ionic state in water), its utility for transmitting an electric current, or its role in accelerating the rusting of iron. A good chemist would probably think of thirty or forty things to do with sodium chloride that I would not have the skill or background to conceive.

## Science Rejects and Excludes the Supernatural from Science

Note that salt can be discussed without invoking God, spiritual qualities of the crystals, or other immaterial aspects that are excluded from science. In one class several years ago, a student brought in some of her mother's crystals (they had been bought at a "new age" store that sold them) and she was describing what each crystal did and what combinations should be worn depending on one's astrological sign. I was not impressed and certainly did not believe that any of this was valid. Yet the student was convinced of its truth on the authority of her mother. To my dismay, most of the students sided with her and felt that I was a stodgy scientist who just had a very limited view of the universe. Unfortunately, this class was not a science course. I was a faculty mentor in a program on "Imagination, Creativity, and Problem Solving." About two-thirds of the students were in the arts and humanities and only a few, mostly premedical students, were in the sciences. I realized that much of humanity, even college-educated people, are very open to supernatural approaches to the universe.

Scientists welcome having their work verified by others as they believe that it will strengthen their work either through extensions of it to related topics or simply in being confirmed. Those who claim they have relics of great religious significance, from pieces of the true cross on which Jesus was crucified to celebrated objects like the shroud of Turin, are more reluctant to have them tested. If they turn out to be hundreds of years younger than they are claimed to be via methods such as carbon-14 dating or tree ring analysis, this would disillusion the faithful who like to believe in the power of such relics in answering their prayers or giving health to the desperate. Many priests shudder when an observant member sees an apparition of the Madonna or Jesus. They fear a faith based on deception or dubious claims. Quite to the contrary are the uncritical acceptances of such visions or the alleged veracity of relics that were acquired in the Middle Ages and thought to be the tears, blood, raiment, or wood associated with the Crucifixion. Any scientist who feared putting their published work through replication would be immediately suspected of fraud. This is why Paul Kammerer's alleged midwife toads with environmentally produced thumb pads took years before they were studied while Kammerer was away. Any scientist who does their work honestly welcomes testing of the data and is eager to assist others in supplying specimens or other materials.

## Science Often Must Work with Incomplete Understanding or Insufficient Data

The more difficult task for science is justification of theories based on incomplete data. Some accounts can never be complete. History is always based on a sampling of the past. Archaeology reconstructs dead cultures with even less evidence than historians have available to them. Biologists can only partially trace the past history of the origins of different groups from one another because so many species have disappeared without a fossil trace of their prior existence. Astronomers can only see so far with their present telescopes and this limits how far back they can go with the alleged Big Bang that produced our present universe. A human ancestor fossil may be represented by a few bones and teeth. These incomplete data make science less exact and more contentious as other scientists rush in with alternative interpretations using the same data. The incompleteness may be due to many factors. Some are intrinsic like the fossil record or the alleged behaviors of people long dead; some are due to the limits of present instruments; some are due to the staggering costs of doing the things needed to provide sufficient data. And yet others are due to complexity itself, which cannot be easily teased apart. A good example in my own field is the discomfort I feel with polygenic traits. There is something a bit strange about large numbers of factors having additive effects. This may be because we think of something like height, longevity, or intelligence as something relatively uniform when it may instead be a composite of so many different processes that what we are measuring are not the effects of many individual genes acting on the same trait but many different genes and environmental factors acting on an immense number of processes that we collectively call height, longevity, or intelligence. When a complex trait is successfully analyzed into its components, as Muller did for the inconstant variable traits, Beaded and Truncate wings, it is no longer a lot of "polygenes" but rather a definable and mappable set of genes that includes a chief gene, different kinds of modifiers that intensify or diminish the expression of the chief gene, and environmental factors (e.g., temperature) that can be controlled along with the number of genetic modifiers used to predict the range and number of flies with different shaped wings in a given cross.

## Science Provides Excitement and Pleasure

Science is also adventure. There is a great deal of excitement and pleasure in field work, collecting specimens, taking measurements, and feeling present in the very nature that you are trying to understand. If you were to ask me about ten years ago what color is the ocean, I would probably have described it as slate gray or blue-green because that is what the sea shore near New York or California looked like. But after going around the world on the *S. S. Universe*, I got to see many seas and how different the oceans look. I was struck by the emerald green of the seas around the Bahamas, the white and gray turbulent waters off the tip of South Africa, and the gorgeous cobalt blue of the Indian Ocean across much of the stretch from Mombasa to Madras (now Chennai). The oceans not only varied in color but also in the waves. What a sight it was to see the oceans with row after row of long billows instead of myriads of tiny waves with white cap sprays. It was also a terrifying sight to see waves large enough to crash on the bow of the ship during a storm and realize that they were several stories tall. For many a geologist or oceanographer, obtaining that information, recording it, measuring it, logging it, and then plotting it onto maps must be as exhilarating as observing it and feeling that thrill of novelty.

There is an altogether different feeling of adventure when the scientist makes voyages of the mind, accumulates data, and sees connections. I can imagine what Fleming or Strassbourger must have felt when they looked at the different plant cells they had stained in the late 1870s, saw the strange cage-like and worm-like masses of chromosomes, and tried to make sense out of them. They did not have the advantage of time lapse photography to reveal the nicely timed stages of cell division, nor could living cells have been observed then for a sufficient length of time to observe chromosomes (vital dyes may not have been sufficient to permit such observations). But as slide after slide showed similar bunching and separations of chromosomes, the first scientists to observe mitosis figured it out. There is a learning curve to almost every activity that leads to discovery or mastery of a technique. I recall my first attempts to cut out and match our twenty-three pairs of human chromosomes. Even with stains that differentiate banding along the arms of the chromosomes, they are not that easy to spot. It takes a few weeks in a cytogenetics laboratory before a new technician gets proficient at spotting the chromosomes, but once they do, it is startling to see them at work and without even needing to cut out the chromosomes, they can spot each chromosome on a photograph. Muller once remarked to us,

"It is no great surprise that we can spot such obvious defects as a bristle that is missing, bent, or forked on a fruit fly. It is no more surprising than spotting a student without a nose around a seminar table." We are so used to seeing human faces that something that obvious would strike us as strange. To all fruit fly workers, even the bristles are familiar. Mastering skills is very much a part of science and it as thrilling to learn for the scientist as any athletic skill for the athlete or artistic skill for the artist.

While scientists pride themselves on how good they are as observers, it is a very limited area of the universe they observe. Being good at finding fruit fly mutations may not transfer to spotting a con artist's sleight of hand that fools the observer. We marvel at street hustlers who do a shuffle of walnut shells and ask us to bet which one contains the penny. Scientists have been fooled by those claiming to do paranormal processes like bending spoons while gently stroking them, but magicians are rarely fooled because they know how to distract and prearrange things, and they can prevent the fraud from performing tricks when under controlled scrutiny.

## Science Relies on Speculation and Inference

One more attribute of science that deserves consideration is speculation. Scientists can be very imaginative in making sense of their data. Many critics of science tell me, "but you're not really a Baconian and you can't induce anything; that's mystical. You must work with some hypothesis and then try to fit the data to prove it." But *every* hypothesis has to come from the imagination! It is just as difficult to explain the creative process of looking at data to come up with a theory as it is to explain the origin of a hypothesis before the data is even collected or examined. Indeed, I can argue that it makes more sense that induction is a better spur of hypotheses than deduction because you have at least had the chance to look at data and gotten some clues as to what may be plausible.

All scientific imagination must be checked against a reality that constantly surprises the scientist. Each disappointment — and there are many — makes the theory change, not the reality. As more attention is paid to a problem and more instruments are devised to probe it, the story gets more complete and older theories melt away. In well-developed fields, it becomes increasingly difficult to add new information. Human gross anatomy is fairly filled with information. The chances of finding a new bone, muscle, or organ is virtually impossible. However, this might not be true for functions. In new fields like those that the human genome project will make available, the potential for discovery, elaborating on theories for the development of organ systems, or figuring out how human behavioral skills are genetically determined from memory to vision, hearing, learning, and imagining may be vast.

If we reflect on science as a whole, we are struck by its diversity, various approaches, unique reliance on an external reality to trim its possible hypotheses, excitement, gift for inventing mental and physical tools, tradition of experimentation, scope for a range of personalities, and respect for accuracy and honesty. Science has its gloomy side too, which includes indifference to consequences, selling out one's values, pettiness of personality defects, and occasional moral lapses. We acknowledge these same flaws in fields as distinct as art and philosophy and attribute them to the diversity of humanity. We must likewise accept the existence of those flaws in science.

I am, of course, a science lover. But I also love art and the humanities, and I recognize my limited abilities in fields outside my one great specialty of genetics. For some 40 years, I have taught science to non-science majors as a humanities course because the humanities give us a sense of who we are and how we are connected to each other as well as the universe. We can celebrate those connections with poetry, music, dance, paintings, sculpting, plays, novels, and essays. Science, too, seizes the imagination and fills our bodies with wonder and the tingle of discovery. Deep within, the scientist is always the child, intrigued and laughing, hoping to grasp the bubble, the feather, or the butterfly.

# 22 How to Live in a Science Saturated World

I have addressed my views of science to three classes of readers. Those who like science already know much of what I have portrayed and whether they are practicing scientists or they just like science, they know how to live with science in their lives. They may be religious, agnostic, or atheist in their outlook on fundamental questions of the human condition. If they are critical of the neglects or abuses of science, they know how to direct their discontentment to bring about change. If change does not occur, they do not feel that they have to abandon their interest in or support for science. Virtually all professional scientists fall into this category, including me. A second category of readers, if I am fortunate to have them read this book, hate science. I used to encounter such students in my Biology 101–102 course at Stony Brook University. I called that course *Biology: A Humanities Approach.* I did so not to placate science haters but to demonstrate to those who knew little about science that my biology course was related to the human condition. It explored the biology we need to know in order to engage in public debates about the abuses of science and the unintended consequences of science. More importantly, it gave another dimension to the human condition. It told us how vulnerable we are to chance events that make some people infertile, some born with birth defects, and some likely to die prematurely of cancer, heart failure, or diabetes. We have no control over how the recombination of chromosomes and genes will turn out in ourselves or our children. My course also taught that the fabric of life at the level of cells, chromosomes, and genes is vulnerable to errors of our own metabolism as well as agents we encounter. Also of importance in seeing biology as part of the humanities is how this knowledge enriches our arts. Can you imagine representing the human face or body without knowledge of anatomy? We see the limits of such an approach in many of the medieval illustrations in manuscripts. They have their own delight in their representations, but would there have been Dürers, Raphaels, or Memlings without the introduction of human anatomy courses for both physicians and artists? Poets and authors sometimes become inspired when they reflect on the findings of science. Our cellularity, molecules, innate inclinations, and developmental unfolding in our mothers' wombs can all enrich the work of a novelist, composer, sculptor, or poet.

The third group of readers will be those who fear science. They have different reasons for their unease. Some do not like the way scientists lend their talents to destructive weapons like atomic bombs that can kill hundreds of thousands of innocent civilians in times of war. Some do not like scientists who produce agents that pollute our waters and atmosphere and occasionally damage our health. They fear science because they see it as intentionally or unintentionally doing considerable harm to humanity. Some fear science because they believe that it is hostile to the arts, humanities, and religion. Evolution is clearly in conflict with the religious beliefs of many students who were taught a view of the universe and human life that is at odds with the ways that science would describe these events. Some students who major in the arts and humanities feel that science is sterile and deprives them of what they call their spirituality or communion with the divine, as well as their belief that nature is holistic and should not be reduced by scientific studies to bits and pieces.

## Concerns About the Bad Outcomes of Sciences

I believe that science should be regulated and to some degree we have a precedent for this. We license physicians and engineers, regulate the introduction of new prescription drugs as well as chemicals that are added to our water supply, foods, or work environment, and require inspections of elevators, bridges, and roads. Such regulation is as important for consumer protection as it is for manufacturers. Products that harm others can lead to costly lawsuits as well as poor publicity, thus scaring away customers. Most manufacturers recognize that without regulation, there is always the risk of rogues who might seek short term gains and ignore the risks of bad outcomes. Regulations can be reviewed and changed as concerns are addressed, as was the case with the introduction of recombinant DNA technology. Those who feared this new approach imagined scenarios of runaway recombinant pathogens being innocently produced through the insertion of genes from other species. However, this never happened. They feared its use in a new kind of germ warfare, but this has not happened either, though it is largely because of an international treaty banning such germ warfare research and making it a war crime to engage in it. As regulation is reviewed after a designated interval of years, the fears of unnecessary regulation or over-restrictive regulation can be addressed.

While some students who initially despised science came to appreciate it more through this approach, some remained unconvinced. This is fine, too. We all recognize that the world is diverse and there are lots of people with different attitudes and preferences. There is much to fear if the nation we live in were forced to share a uniform vision of religion, social policy, and political outlook. Where such societies do exist, we often deplore the lack of freedom or oppression of those who reject the authority of the state in enforcing such obedience. Environmental concerns are more difficult to assess. There are competing values when those in the forestry industry contend with those who see nature as, if not sacred, at least being vulnerable to exploitation by commercial needs. Most foresters will replant seedlings to make sure that there will be trees to cut decades from now. How much of a mature forest should be opened to commercial logging is very difficult to assess because the question is usually treated not as a scientific, but as a political issue. There is less controversy about overfishing in places where large populations of commercially useful fish abound. Many species over the decades have disappeared or a form of natural selection has led to the disappearance of larger sized fish species. Where small business fisheries form organizations, they tend to regulate themselves, but when two or more nations use the same open waters for fishing, it can be disastrous to the survival of fish species. Some exploitation of animals is based on human ignorance. There is nothing in a rhinoceros horn that will cure impotence in older men, and yet the rhinoceros is on the brink of extinction because of this ancient fallacy. Other exploitations are based on commercial value, like elephant tusks for the ivory art that skilled artisans create. Poachers who continue to bribe officials or take risks to kill elephants either do not see that they are contributing to the extinction of these mammals or believe that this is one of the few ways that they can earn a decent living. In the United States, hunting was unregulated when land was abundant, the population was sparse, and hunting was necessary to feed one's family. Today, hunting is regulated and used to maintain acceptable numbers of a given species in the wild. For most hunters, it is a sport. Still controversial is the hunting of wolves that are seen as a threat by ranchers raising livestock.

## Religion and Science Have Many Conflicts

It is not just evolution that leads to conflicts with religion. Religions have different rules about behavior, many of which deal with sexual reproduction and sexual identity. A third feature of the conflict deals with the issue of morality or ethical behavior. We will look at all three of these and explore what can be done to lessen these conflicts. In the United States, this is particularly acute because we have more denominations of religions and more people who identify as religious compared to most other industrial nations. For the evolution controversy, the courts have consistently kept religious interpretations of the origins of life and the age of the universe out of public school science courses. The debate centers on those who see the Bible as dictated to scribes by God and thus inerrant. Since 1910 when the fundamentalist movement was founded, this interpretation has spread and those who follow it have been the major leaders in lobbying school boards and state legislatures to adopt one of several policies. The first was the attempt to ban the teaching of evolution in science classes, which was defeated in the 1920s with bad publicity for fundamentalism at the Scopes trial in Dayton, Tennessee. The second approach was to argue for equal airtime for creationist interpretations of the universe and life on earth, which was defeated in the 1950s by a unanimous Supreme Court ruling that held it to be a violation of the separation of church and state. Attempts to get around this by claiming that it is not biblical creation but "creation science" that is offered as a legitimate science have also been struck down by the courts as a stealth approach to bringing religion into the science classroom. The third approach has been the influence of the fundamentalist lobby on the purchase or content of school textbooks that teach evolution or, if they do, that exclude creationist views in science texts. This leads to watered down versions of science texts where the age of the earth or the universe is omitted, as are cosmology theories and any implications of a past evolution of life. Such encounters will continue in the United States for the foreseeable future. Most Catholics, Protestants, and Jews do not have a problem with evolution and use the argument that God used scientific methods to bring about the universe and life on earth, not miracles of sudden creation.

## Reproduction and Religion are Sometimes at Odds

The regulation of reproduction was primarily a Roman Catholic concern before the 1970s. Protestant involvement was parallel and more diverse. In the United States, anti-Catholic hostility among fundamentalists was predominant until then. Catholic views on reproduction are based on a male and female model of human sexuality. Anything other than heterosexual reproduction by vaginal intercourse during marriage was considered a violation of natural law or contradictory to Church teachings and thus immoral. For this reason, contraception with few exceptions was prohibited by the Church. The birth control movement, initiated by Margaret Sanger and Marie Stopes in the early twentieth century, changed how women in industrialized nations had children. They wanted family planning so that they could devote more time to their children, have time for their own education and personal lives, work if they had the skills to use them, and care for their own health. Eventually the Comstock laws of the 1870s, which banned sex education by physicians or other professionals as well as the prescription and use of contraceptives, were nullified. The Comstock laws were promoted by New England Puritan descendants who saw almost all sexual activity as sinful. Their critics called them "blue noses." As the birth control movement swept across the industrialized world, the two-child family became the norm and family planning was held as a virtue rather than a vice. In the 1970s, this more relaxed view of the privacy of reproductive decisions was extended to abortion in the *Roe v. Wade* decision, unleashing a battle that continues into the twenty-first century. As same sex activity was decriminalized by the courts and legislation, the alliance of Catholic and fundamentalist religions has emerged in their rejection of what they see as criminal activity (abortion) or a violation of God's mandate (consenting adult sex that violate moral teachings). Science has illuminated the workings of the reproductive system and the ways it can be hormonally regulated, and produced the devices that can be used to prevent or stop fertilization, implantation, or the initial stages of embryonic development. Science introduced the new field of IVF or in vitro fertilization to assist the infertile so that they can have children. Science has also challenged simplistic views of life such as the establishment of a person or soul at fertilization. This view is problematic with regards to identical twins, of which some 80 percent arise after implantation of the blastocyst and the formation of an inner mass of stem cells. If souls are not shared or split into two with twinning, then they cannot be initiated during zygote formation for these future twins. Such arguments infuriate those who hold that life begins with fertilization, although the recognition that fertilization involves the union of one sperm with one egg did not happen until the 1870s. Nor does the contradictory use of the term conception bother those who hold this religious view. The term conception was originally associated with the implantation of the embryo into the uterus. The idea that fertilization either existed or took place in the oviduct was not known when most religious rules or laws were set. We do not know how these religious conflicts with the reproductive technology and practices of today's adults will turn out. Funding for health, especially women's health, is often involved because it is usually the female who chooses the means of family planning in most families. Science could change the situation if a male hormonal or other chemical means were worked out to prevent spermatogenesis and was available by prescription or as an over the counter medication like aspirin. This would eliminate the association of several female contraceptives with abortion, but it would still be opposed by Catholic teaching as against what they call natural law and its critics call Catholic theology.

# The Insecurity of Faith in the Presence of Science

Much more difficult to assess are the claims that science diminishes faith in children exposed to it; that it destroys the human spirit (in its broadest sense of concern about the major philosophic questions about the meaning of life); that it leads to amoral or immoral behavior; and that it leaves humanity with a void that cannot be filled by a shallow materialism. I do not doubt that there are people who feel that this has happened to them. One could argue that the opposite happens when people are attracted to cults (or very demanding religions) and everything except fellowship and obedience to the cult's teachings melt away to the dismay of relatives and friends who are abandoned.

I do not believe I am alone in my enjoyment of life as a scientist with a happy family life, friends, appreciation by many students and colleagues, and the satisfactions of an examined life. I was raised as an atheist and remain an atheist in my belief that I have no need for a concept of god or any supernatural belief system. I am among those who are lucky to have experienced a long life (now going into my late 80s). I try to be an ethical and moral person. Kant's empathy test of ethical behavior works for me. He noted that if a sane healthy adult is asked if he or she wants to be murdered, he or she would reply no. Sane adults wants to live; there are no exceptions. One could say the same about being robbed, assaulted, raped, or lied to. Sane people do not wish for such things to happen to anyone and certainly not to themselves. This logic has led Kant to argue that it is universal that we should not kill, maim, rob, or lie to others. One can have universal moral laws without their being provided by God on a set of stone tablets. It is false to believe that a person who does not believe in God will go on a binge of immoral behavior because there is no fear of punishment. Many people help others, feel compassion for those who are victims of fate or bad social policy, or enjoy dealing honestly with others because they feel great satisfaction in doing so, not because they fear punishment if they fail to do so. At the same time, I consider myself religious. I have been a Unitarian since 1960. It is a religion that has no creed. It fosters individual search for meaning and its members range from theists to atheists. What makes it a religion is its fellowship and belief that humans can help others and make the world better in each coming generation. One does not need to believe in a God to do that.

More debatable is the way we construct our lives. I believe that happy people can be religious or non-religious. Our personalities differ whether there are innate factors involved or not. Some people who are religious are not happy and they can cause a lot of misery to others through their carping personalities. There are those for whom a life without an afterlife would be meaningless. Others are epicurean in their outlook and feel that life is a gift and that they should make good use of it, not because they are told to do so but because it is how their personality works. Living well is a challenge because there are many ways to live a satisfying life. Some people triumph over their infirmities and insecurities, some find deep pleasure in their talents or their careers, and some are rewarded by the closeness of their family members. I do not doubt that there were happy people in some of the unhappiest eras to be alive.

Without science, we would have a very different world with no germ theory to prevent infectious diseases, no technology to prevent periodic starvation from famines, limited ways to keep our bodies clothed against harsh winter weather, high infant mortality, and considerably shorter life expectancy. We must remind ourselves of the good that science has provided in giving us some control over our lives and environments, increasing the opportunities for education, making musical instruments that generate musical creativity, making deep insights into how our bodies work, revealing the 92 elements out of which the material universe is composed, explaining how clouds form, volcanoes erupt, and earthquakes happen, giving us tools to explore the universe, opening up unseen galaxies and their myriads of stars, and allowing us to see our own cellularity and follow our life cycle from fertilization to birth and death. There is no foreseeable closure to this repetition of the daily rhythm of the earth's rotation, generations past and future, and science providing ever new insights into our universe and the pleasure of living in it.

# References to Works Cited

Allen G (1978) *Thomas Hunt Morgan: The Man and His Science*. Princeton University Press, Princeton, NJ.

Altenburg E *T. H. Morgan, Democrat*, Manuscript, Lilly Library, archives.i.u.edu>-catalog.InU-Li-VADG19.

Bacon F (1629) *Novum Organum*. Collier and Sons, New York, NY. (English translation) 2014.

Berg P, Singer M (1995) The recombinant DNA controversy: Twenty years later. *Proc Natl Acad Sci U S A* **92**: 9011–9013.

Blunt W (1971) *The Compleat Naturalist: A Life of Linnaeus*. Viking Press, New York, NY.

Carlson E (2018) *How Scientific Progress Occurs: Incrementalism and the Life Sciences*. Cold Spring Harbor Laboratory Press, Cold Spring Harbor, NY.

Carlson E (2001) *The Unfit: A History of a Bad Idea*. Cold Spring Harbor Laboratory Press, Cold Spring Harbor, NY.

Carlson E (1981) *Genes, Radiation, and Society: The Life and Work of H. J. Muller*. Cornell University Press, Ithaca, NY.

Churchill FB (2015) *August Weismann: Development, Heredity, and Evolution*. Harvard University Press, Cambridge, MA.

Codesco F (1973) *The Shaping of American Graduate Education: Daniel Coit Gilman and the Protean PhD*. Rowman and Littlefield, Totowa, NY.

Crick F (1958) On protein synthesis. *Symp Soc Exp Biol* **12**: 138–163.

Darwin C (1859) *The Origin of Species*. John Murray, London, UK.

Darwin C (1839) *Voyage of the Beagle*. Henry Colburn, London, UK. http://www.bl.uk/manuscripts/FullDisplay.aspx?ref=Arundel_MS_263.

Da Vinci, Leonardo *Notebooks Codex Arundel (1480–1518)* British Museum (on-line) Turning the Pages.

Dickens C (1854) *Hard Times*. Bradbury and Evans, London, UK.

Drake S (1990) *Galileo: Pioneer Scientist*. University of Toronto Press, Toronto, CA.

Godwin W (1973) *Enquiry Concerning Political Justice and Influence on General Virtue and Happiness*. G. G. and J. Robinson, London, UK.

Goethe JW. *Faust part I (1806) and Part II (1831)* in English Gutenberg Project.

Gross PR, Levitt N (1994) *Higher Superstition: The Academic Left and its Quarrel with Science*. Johns Hopkins University Press, Baltimore, MD.

Hardy GH (1940) *A Mathematician's Apology*. Cambridge University Press, Cambridge, UK.

Hardy GH (1908) Mendelian proportions in a mixed population. *Science* **28**: 49–50.

Hixson JR (1976) *The Patchwork Mouse: Politics and Intrigue in the Campaign to Conquer Cancer*. Anchor Doubleday, New York, NY.

Hooke R (1665) *Micrographia, or, Some Physiological Descriptions of Minute Bodies Made by Magnifying Glasses: With Observations and Inquiries Thereupon*. J. Martyn and J Allestry, London, UK.

Hughes A (1959) *A History of Cytology*. Abelard-Schuman, London, UK.

Huxley A (1932) *Brave New World*. Chatto and Windus, London, UK.

Iltis H (1932) *Life of Mendel*. W. W. Norton, New York, NY.

Keller EF (1983) *A Feeling for the Organism: The Life and Work of Barbara McClintock*. Freeman, New York, NY.

Koestler A (1972) *The Case of the Midwife Toad*. Random House, New York, NY.

Lewis S (1925) *Arrowsmith*. Harcourt Brace, New York, NY.

Linnaeus C *Systema Naturae* (1735) facsimile of the first edition with an introduction and a first English translation, Nieuwkoop, Holland: De Graaf, 1964.

Lock RH (1906) *Recent Progress in the Study of Variation, Heredity, and Evolution*. J. Murray, London, UK.

Lyons SL (2020) *From Cells to Organisms: Re-envisioning Cell Theory* University of Toronto Press, Toronto.

Malthus T (1798) *An Essay on the Principle of the Human Population*. J. Johnson, London, UK.

Mann T (1924) *Der Zauberberg*. Fischer, Berlin, DE. English version *The Magic Mountain* 1994 translation by J Edwards, Vintage Books.

Marlowe, Christopher *The Tragical History of the Life and Death of Doctor Faustus Quarto 1604* Gutenberg Project, Ebook#779 November 3, 2009.

McLaren A (1976) *Mammalian Chimaeras*. Cambridge University Press, New York, NY.

Medvedev ZA (1969) *The Rise and Fall of T. D. Lysenko.* Columbia University Press, New York, NY.

Mintz B (1964) Formation of genetically mosaic mouse embryos and early development of lethal (t12/t12)-normal mosaics. *J Exp Zool* **157**: 267–272.

Muller HJ (1936) Physics in the attack on the fundamental problems of genetics. *Sci Mon* **44**: 210–214.

Muller HJ (1929) The gene as the basis of life. Read before Int. Congr. Plant Sci., Ithaca, Aug. 1926. *Publ Proc Int Congr Plant Sci* **1**: 897–921.

Muller HJ (1950) Our load of mutations. *Am J Hum Genet* **2**(2): 111–176.

Muller HJ (1927) The problem of genic modification. Presented in Berlin, 5th Genetic Congress 1927 Proceedings of the Fifth International Congress of Genetics, Berlin 1927 *Zeitschrift für inductive Abstammungs- und Vererbungs-lehre* Suppl. 1: 234–260.

Muller HJ (1922) Variation due to mutation in the individual gene. *Am Nat* **556**: 32–50.

*New York Times* April 20, 1949 page 1 column 2 Misleading the public on radiation via the Atomic Energy Commission and President Eisenhower.

Nirenberg MW, Matthaei JH (1961) The dependence of cell-free protein synthesis in *E. coli* upon normally occurring or synthetic polyribonucleotides. *Proc Natl Acad Sci U S A* **47**: 1588–1602.

Oppenheimer JM (1970) Hans Driesch and the theory and practice of embryonic transplantation. *Bull Hist Med* **44**(4): 378–382.

Paley W (1802) *Natural Theology: Or, Evidences of the Existence and Attributes of the Deity.* J. Faulder, London, UK.

Pires AM, Branco JA (2010) A statistical model to explain the Mendel-Fisher controversy. *Stat Sci* **25**: 545–565.

Plotz D (2005) *The Genius Factory: The Curious History of the Nobel Prize Sperm Bank.* Random House, New York, NY.

Poe EA (1848) *Eureka: A Prose Poem.* G. P. Putnam, New York, NY.

Popper K (1962) *Conjectures and Refutations: The Growth of Scientific Knowledge.* Basic Books, New York, NY.

Saap J (1990) *Where the Truth Lies: Franz Moewus and the Origins of Molecular Biology.* Cambridge University Press, New York, NY.

Sagan C (1973) *The Cosmic Connection: An Extraterrestrial Perspective.* Anchor Press/Doubleday, New York, NY.

Schrödinger E (1944) *What is Life?: The Physical Aspect of the Living Cell.* Cambridge University Press, New York, NY.

Shelley MW (1818) *Frankenstein, or the Modern Prometheus.* Lackington, Hughes, Harding, Mavor, & Jones, London, UK.

Shurkin JN (1992) *Terman's Kids: The Groundbreaking Study of How the Gifted Grow Up.* Little, Brown & Co, Boston, MA.

Skinner BF (1948) *Walden Two.* Macmillan, New York, NY.

Snow CP (1959) *The Two Cultures and the Scientific Revolution.* Cambridge University Press, New York, NY.

Timofeev-Ressovsky NW, Zimmer KG, Delbrück M (1935) Über die natur der genmutation und der genstruktur. *Nachr Ges Wiss Göttingen FG VI Biol N F* **13**: 189–245.

Turgenev I (1862) *Fathers and Sons.* English translation Project Gutenberg, December 21, 2009 E Book # 30723.

von Humboldt A (1845) *Cosmos: A Sketch of a Physical Description of the Universe.* Longman, London, UK.

Vonnegut K (1963) *Cat's Cradle.* Holt, Rinehart and Winston, New York, NY.

Watson JD (1968) *The Double Helix.* Athenaeum, New York, NY.

Weinberg W (1908) Über den Nachweis der Vererbung beim Menschen. *Jahreshefte Vereins vaterlandische Natur, Wurtemberg* **64**: 369–382.

Weismann A (1891) The continuity of the germplasm as the foundation of theory of heredity. In A Shipley (ed), *Essays Upon Heredity and Kindred Problems.* pp.163–255 Oxford University Press, Oxford.

Wilson EB (1896) *The Cell in Development and Inheritance.* Macmillan, New York, NY.

# About the Author

Elof Axel Carlson was born in Brooklyn, NY in 1931 and attended NYU as an undergraduate (BA 1953) and Indiana University for his doctoral studies (PhD 1958). His dissertation work was with Nobel geneticist H. J. Muller. Carlson has taught at Queen's University (Canada), UCLA, and Stony Brook University in New York's Long Island. He is now retired and lives in Bloomington, Indiana as a Visiting Scholar in the Institute for Advanced Study of Indiana University. Carlson is married to Nedra (née Miller) who was an in vitro fertilization embryologist and they have five children, 12 grandchildren, and three great-grandchildren. Carlson is the author of ten

Photo by Village Times Herald 1997

published books and four edited books. He was recipient of the Harbison Award for Gifted Teaching by the Danforth Foundation and he was elected a Fellow of the American Association for the Advancement of Science in 1963. He writes a newspaper column, *Life Lines*, which appears in the *Times-Beacon-Record* newspapers of Long Island publisher Leah Dunaief. Carlson is a geneticist and historian of science who has enjoyed a career of learning, teaching, and writing.

## Books by Elof Axel Carlson

1. *The Gene: A Critical History*
2. *Genes, Radiation, and Society: The Life and Work of H. J. Muller*
3. *The Unfit: A History of a bad Idea*
4. *Mendel's Legacy: The Origin of Classical Genetics*
5. *Times of Triumph, Times of Doubt: Science and the Battle for Public Trust*
6. *Neither Gods nor Beasts: How Science is Changing Who We Think We Are*
7. *Mutation: The History of an Idea from Darwin to Genomics*
8. *The 7 Sexes: Biology of Sex Determination*
9. *How Scientific Progress Occurs: Incrementalism and the Life Sciences*
10. *How to Construct Your Intellectual Pedigree: A History of Mentoring in the Life Sciences*

# Index

Printed in the United States
by Baker & Taylor Publisher Services

Printed in the United States
by Baker & Taylor Publisher Services